Springer Series on Atomic, Optical, and Plasma Physics

Volume 119

The Springer Series on Atomic, Optical, and Plasma Physics covers in a comprehensive manner theory and experiment in the entire field of atoms and molecules and their interaction with electromagnetic radiation. Books in the series provide a rich source of new ideas and techniques with wide applications in fields such as chemistry, materials science, astrophysics, surface science, plasma technology, advanced optics, aeronomy, and engineering. Laser physics is a particular connecting theme that has provided much of the continuing impetus for new developments in the field, such as quantum computation and Bose-Einstein condensation. The purpose of the series is to cover the gap between standard undergraduate textbooks and the research literature with emphasis on the fundamental ideas, methods, techniques, and results in the field.

More information about this series at https://link.springer.com/bookseries/

John Rice

Driven Rotation, Self-Generated Flow, and Momentum Transport in Tokamak Plasmas

Springer

John Rice
Plasma Science and Fusion Center
Massachusetts Institute of Technology
Cambridge, MA, USA

ISSN 1615-5653 ISSN 2197-6791 (electronic)
Springer Series on Atomic, Optical, and Plasma Physics
ISBN 978-3-030-92265-8 ISBN 978-3-030-92266-5 (eBook)
https://doi.org/10.1007/978-3-030-92266-5

This Springer imprint is published by the registered company Springer Nature Switzerland AG
The registered company address is: Gewerbestrasse 11, 6330 Cham, Switzerland

Preface

Examples of rotation can be found everywhere in nature, from water circling the drain to hurricanes to spots on Jupiter to entire galaxies. Rotation is also seen in high temperature plasma experiments, often with no obvious cause. The purpose of the present study is to document and attempt to understand the experimentally observed rotation in tokamaks, toroidal (donut shaped) devices used to contain high temperature (a few keV $\approx$ 10 s of millions of degrees) plasmas. While there are several external momentum sources which can be used to drive rotation in tokamaks, spontaneous, self-generated flow, up to ~100 km/s, is routinely observed in plasmas without external momentum input. Furthermore, this intrinsic rotation can reverse direction in a fraction of a second without any obvious reason. The purpose of tokamak research is to provide an energy source *via* controlled thermonuclear fusion, and in order to achieve this in a cost effective manner, it is important to understand the rich variety of phenomena in tokamak plasmas. What better opportunity is there than explaining self-generated flow?

The influence of rotation on tokamak performance has been widely demonstrated. For instance rotation can stabilize certain deleterious magnetohydrodynamic (MHD) instabilities, and the rotation gradient has been shown to suppress turbulence, leading to transport barriers and an enhancement in energy confinement. In this regard it is important to understand the rotation in tokamak plasmas, and to use this understanding to control the rotation spatial profile. There are several recent experimental [1–5] and theoretical [6–10] surveys on this subject, and the present goal is to update and expand upon the observational reviews.

Toroidal velocity profile evolution is governed by a balance between external torques (momentum sources and sinks) and the momentum flux gradient [4] as

$$m \cdot \partial(n v_\phi)/\partial \mathrm{t} = -\nabla \cdot \Gamma_\phi + S_{\mathrm{ext}} \qquad (1)$$

where v_ϕ is the toroidal velocity, n is the density, m is the particle mass, Γ_ϕ is the momentum flux and S_{ext} includes external momentum sources and sinks. There are several momentum sources, e.g. direct drive from neutral beam injection or radio frequency waves, and indirect drive from the electric field created by orbit shifts and losses. Sinks include neutral damping and magnetic braking (neo-classical toroidal viscosity). Sinks and indirect sources depend upon the ambient plasma parameters. The magnetic geometry and subsequent particle orbits must be considered since non-ambipolarity will give rise to an electric field which enters into the force balance and collisions must also be taken into account; these are the ingredients for neo-classical theory.

The momentum flux is dominated by the toroidal Reynolds stress [11] which consists of three parts:

$$\Gamma_{\phi}/\mathrm{m} = -\chi_{\phi}\partial \mathrm{v}_{\phi}/\partial \mathrm{r} + \mathrm{V}_{p}\mathrm{v}_{\phi} + \Pi^{\mathrm{res}} \tag{2}$$

where χ_{ϕ} is the momentum diffusivity (or viscosity), V_p is the momentum pinch (or convection) and Π^{res} is the residual stress. All three of these transport coefficients can be dominated by different competing effects which also depend upon ambient plasma parameters. The latter component, which is independent of both the toroidal velocity v_{ϕ} and its spatial gradient $\partial \mathrm{v}_{\phi}/\partial \mathrm{r}$, has the unique characteristic that it can function as a momentum source (and so appears on both sides of the force balance), with the intrinsic torque density given by $\nabla \cdot \Pi^{\mathrm{res}}$ [11]. Π^{res} is governed by turbulence which is often driven by spatial gradients of plasma parameters. Following a summary of rotation velocity measurement techniques (Chap. 1) will be an examination of momentum sources (Chap. 2) and sinks (Chap. 3), comparisons with the predictions of neo-classical theory (Chap. 4), a detailed discussion of the residual stress as a momentum source (Chap. 5) and a review of the momentum transport coefficients χ_{ϕ}, V_p and Π^{res} (Chap. 6).

It is important to understand all aspects of this problem, covered in Chaps. 2–6. Each of these effects has a different plasma parameter dependence (such as density, temperature, neutral density, collisionality) and each can dominate in different regions of operational space. It is necessary to consider magnetic geometry, particle orbits, collisions, electric fields, MHD constraints through current and pressure gradients and the influence of turbulence through density, temperature, pressure and current density gradients. Further complicating this problem is that the velocity is a vector, with direction and magnitude. The challenge is to account for this myriad of effects in a comprehensive manner. This process will be systematically addressed as follows, after a discussion of velocity measurement techniques.

The most common method of velocity determination, from Doppler shifts of atomic transitions, will be covered in Sect. 1.1, with a review of passive (Sect. 1.1.1) and active (Sect. 1.1.2) techniques. Direct external momentum sources will be considered in Sect. 2.1, including neutral beam injection (NBI) and various radio frequency schemes (ICRF, LH and ECH). In Sect. 2.2, indirect rotation drive due to $\mathbf{j}\times\mathbf{B}$ forces arising from radial orbit shifts and non-ambipolar effects, such as toroidal magnetic field ripple loss, will be reviewed. A summary of momentum sinks will be presented in Chap. 3, including neutral damping, magnetic field perturbations and neo-classical toroidal viscosity. In Chap. 4 comparisons of observations with neo-classical calculations will be shown, including both poloidal and toroidal rotation. Self-generated flow arising from the residual stress will be discussed in great detail in Chap. 5. Two general categories of intrinsic rotation will be examined, that occuring in the enhanced confinement regimes (H- and I-mode) and that in L-mode plasmas, including the curious rotation reversal phenomenon (and its connexion with confinement saturation and "non-local" heat transport cut-off). Chapter 6 will be concerned with determination of momentum transport coefficients: the momentum diffusivity, the momentum pinch and the residual stress. A discussion of open questions will be the subject of Chap. 7. In the following, the toroidal rotation will be written as v_ϕ or V_{Tor} while the poloidal rotation will be denoted as v_θ or V_{Pol}. A majority of velocity measurements is of impurity rotation, which is often used as a proxy for main ion rotation.

Cambridge, MA, USA John Rice

Acknowledgments

Enlightening interactions with Clemente Angioni, Manfred Bitter, Yann Camenen, Norman Cao, Jonathan Citrin, John deGrassie, Peter deVries, Patrick Diamond, Basil Duval, Lars-Goran Eriksson, Chi Gao, Martin Greenwald, Brian Grierson, Jerry Hughes, Katsumi Ida, Alex Ince-Cushman, Davis Lee, Earl Marmar, Rachael McDermott, Yong-Su Na, Arthur Peeters, Yuri Podpaly, Thomas Pütterich, Matt Reinke, Timothy Stoltzfus-Dueck, Tuomas Tala, Steve Wolfe, and Maiko Yoshida are gratefully acknowledged. Work supported at MIT by DoE Contract No. DE-FC02-99ER54512.

Contents

Chapter 1
Velocity Measurements in Tokamaks

In order to study rotation it is necessary first to measure it reliably. There are several techniques in use on tokamaks which will be summarized anon: Doppler shifts of atomic transitions, MHD mode rotation from frequency analysis, probe measurements and from microwave scattering. For the atomic transitions, the population of the upper levels can either be passive (*via* collisional excitation or radiative recombination from plasma electrons) or active (from charge exchange recombination with injected neutrals).

1.1 Doppler Shifts of Atomic Transitions

The most common method of measuring the rotation is from observation of the Doppler shifts of atomic transitions, either in impurities (intrinsic or extrinsic) or background ions (hydrogen, deuterium and/or helium). Depending upon the appropriate plasma conditions, these measurements can be made in the visible, ultraviolet (UV) or x-ray wavelength ranges.

1.1.1 Passive Spectroscopy

Early velocity determination in tokamaks was from passive measurements of intrinsic impurity ions whose upper levels were populated by collisional excitation. The first velocity measurement, in the VUV wavelength range, was from O V (O^{5+}) emission in the LT-3 device [12]. Similar observations were subsequently made on PLT [13, 14], Torus II [15], PDX [16], TM-4 [17], JET [18], TCA [19], COMPASS-D, TEXT [20] and TCABR [21, 22]. Main ion velocities have also been inferred from observations of D_α light in AUG [23]. The PLT measurements

J. Rice, *Driven Rotation, Self-Generated Flow, and Momentum Transport in Tokamak Plasmas*, Springer Series on Atomic, Optical, and Plasma Physics 119,
https://doi.org/10.1007/978-3-030-92266-5_1

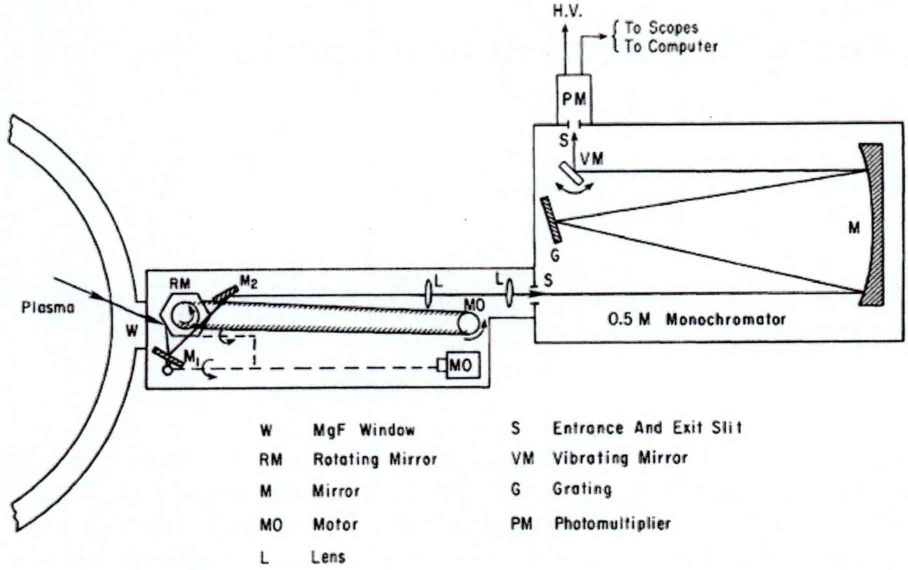

Fig. 1.1 Schematic of instrumentation for measurement of poloidal or toroidal intensity distributions with simultaneous spectral scan of an emitted line. Reprinted with permission from Fig. 1 of [13] S. Suckewer et al. 'Toroidal Plasma Rotation in the Princeton Large Torus Induced by Neutral-Beam Injection', Phys. Rev. Lett. **43**, 207 (1979). Copyright (1979) by the American Physical Society

[13] were from a clever spectrometer system, with a rotating mirror to sample the plasma cross section and a fast vibrating mirror to scan the individual line shapes, a layout overview of which is shown in Fig. 1.1. To demonstrate the capabilities of this spectrometer, time histories of the rotation velocity during an NBI heated PLT plasma from this system are shown in Fig. 1.2. In the core of this $T_e \sim 2$ keV discharge, the toroidal rotation reached a level in excess of 100 km/s, as determined from the Doppler shift of a forbidden Fe XX (Fe^{19+}) transition at 2665 Å. Neutral beam power was applied between 350 and 500 ms. Emission lines from other ions (C V and H I), which fell in the spectral range of the spectrometer show lower velocity from the peripheral regions of the plasma. A radial velocity profile concocted from these three separate emission lines during the NBI phase for this plasma is shown in Fig. 1.3. In this case the toroidal rotation profile was strongly peaked near the plasma center.

One drawback of using VUV and visible emission lines (whose upper levels are populated by collisional excitation) is that they are typically from impurity charge states that exist in narrow spatial shells, thereby only allowing access to one location in the plasma. While it is possible to cobble together a complete radial profile using a variety of different impurities and charge states [13, 18] (as in Fig. 1.3), it would be desirable to utilize closed shell ionization states (He-like or Ne-like) which exist over a larger spatial region. Another limitation of VUV and visible transitions is

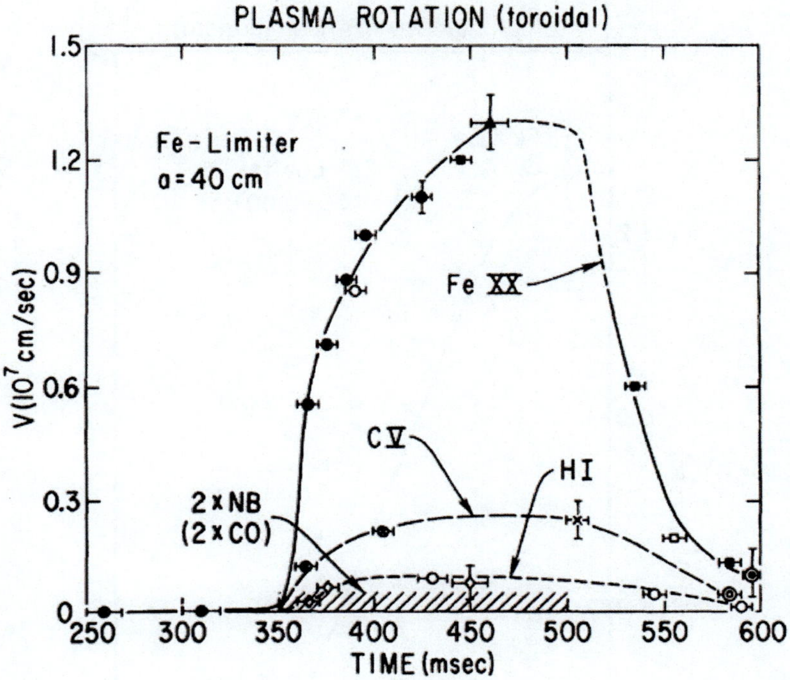

Fig. 1.2 Time evolution of the toroidal plasma velocity induced by two co-injected neutral beams, with total power 1.5 MW, in plasma with central electron temperature $T_e(0) \approx 2$ keV and density $n_e(0) \approx 3$–$5 \times 10^{19}/m^3$. The different atoms refer to different radial locations in the plasma. Reprinted with permission from Fig. 3 of [13] S. Suckewer et al. 'Toroidal Plasma Rotation in the Princeton Large Torus Induced by Neutral-Beam Injection', Phys. Rev. Lett. **43**, 207 (1979). Copyright (1979) by the American Physical Society

that they usually sample regions of the plasma with lower electron temperature (a notable exception is the forbidden Fe XX transition). These two shortcomings can be ameliorated by using x-ray transitions from He-like ions. The first x-ray spectrometer in the Johann configuration [24] was installed on PLT for viewing He-like iron [25]. A schematic of this system is shown in Fig. 1.4. Such a spectrometer requires a large area beryllium window, a carefully bent crystal and a linear position sensitive detector. In the early manifestations of these spectrometers, various types of proportional counters were employed. Johann x-ray spectrometers have been used on PDX [16], TFTR [26, 27], DIII-D [28], JET [18, 29], Tore Supra [30], FT and JIPP-TII. An example of a velocity measurement using the He-like titanium (Ti XXI or Ti^{20+}) resonance line at 2.6097 Å from a neutral beam heated TFTR plasma [26] is shown in Fig. 1.5. During the NBI pulse from 2.3 to 2.8 s, the core toroidal velocity reached 150 km/s. One drawback of Johann spectrometers is that in order to achieve high wavelength resolving power, the radius of the Rowland circle, and hence the system size, becomes large. An alternative is to use the von Hamos configuration [31], which allows for a compact spectrometer. The layout for

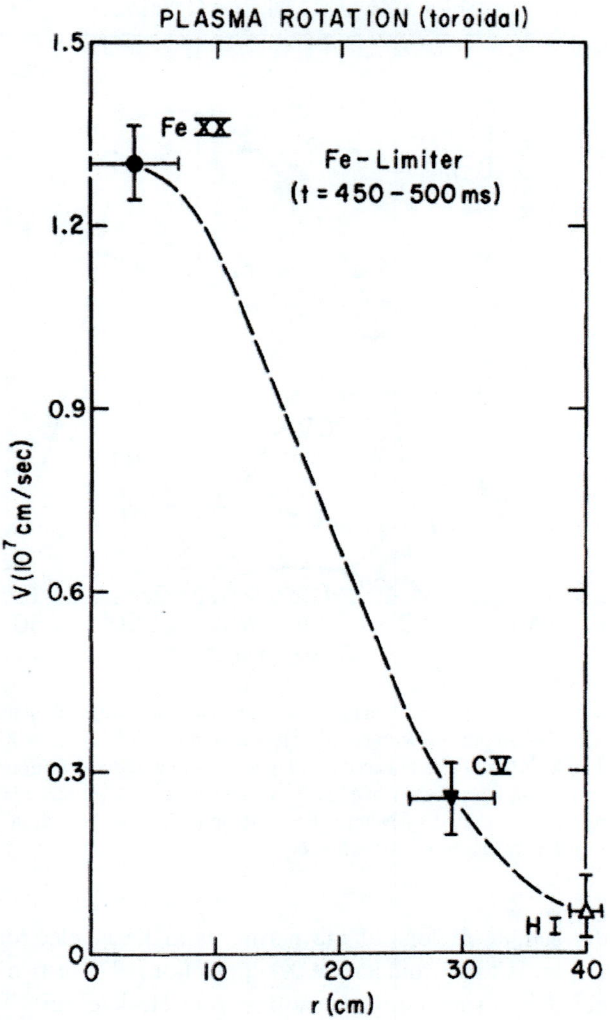

Fig. 1.3 Radial distribution of toroidal plasma velocity at 100–140 ms after beginning of neutral beam injection. Reprinted with permission from Fig. 4 of [13] S. Suckewer et al. 'Toroidal Plasma Rotation in the Princeton Large Torus Induced by Neutral-Beam Injection', Phys. Rev. Lett. **43**, 207 (1979). Copyright (1979) by the American Physical Society

the von Hamos system of Alcator C [32] is shown in Fig. 1.6. The cylindrically bent crystal is flat in the dispersive plane and unlike the Johann system, an entrance slit is required for spatially extended sources. This allows for a small area Be window. Von Hamos type spectrometers have also been implemented on Alcator C-Mod [33] and an example of Doppler shifted spectra [34] from H-like Ar^{17+} (extrinsically introduced) and Ne-like Mo^{32+} (intrinsic) is shown in Fig. 1.7. These spectra were

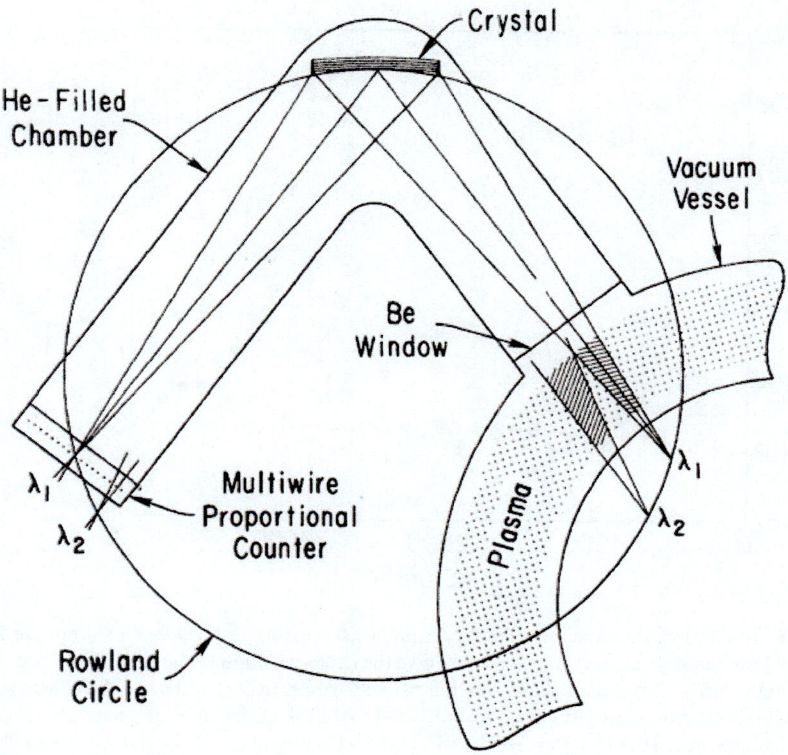

Fig. 1.4 Quartz crystal spectrometer in the Johann configuration. X-rays emitted from the shaded area of the plasma pass through a beryllium window into a helium filled tube and are Bragg reflected by the crystal onto the position sensitive detector. Photons of different energies are focussed to different detector points. Reprinted with permission from Fig. 1 of [25] M. Bitter et al. 'Doppler-Broadening Measurements of X-Ray Lines for Determination of the Ion Temperature in Tokamak Plasmas', Phys. Rev. Lett. **42**, 304 (1979). Copyright (1979) by the American Physical Society

from two Ohmic L-mode discharges with the plasma current in opposite directions and exhibit counter-current rotation with magnitudes of ~40 km/s.

While it is possible to obtain spatial velocity profiles by utilizing an array of the individual spectrometers just described [33, 35, 36], a more elegant approach is to use a spatially imaging system. This can be achieved with a spherically bent crystal in the Johann configuration [37, 38] and a schematic of a spectrometer system is shown in Fig. 1.8. This arrangement requires a 2D x-ray detector: one direction for dispersion (meridional plane) and the other for spatial imaging (sagittal plane). Such a system has been implemented on Alcator C-Mod [39] and the layout is shown in Fig. 1.9. There are actually two spectrometers in the same housing, one for viewing He-like argon and the other for H-like. The He-like argon system has complete up/down coverage of the plasma cross section, as seen in the bottom frame. Measurements with this system are chord averaged, so a tomographic inversion

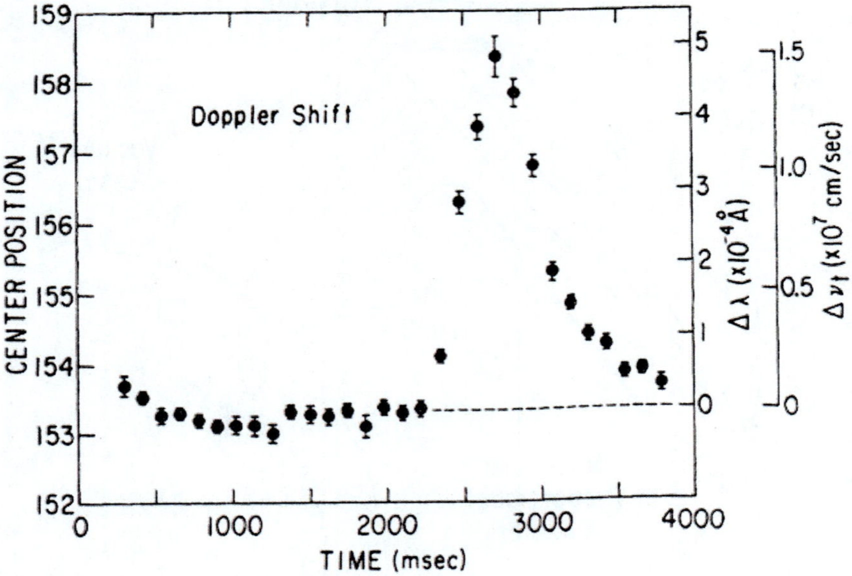

Fig. 1.5 Toroidal plasma rotation velocities from measurements of the Doppler shift of the Ti XXI K_α line. Left and right scales on the ordinate give the center position of the Ti XXI K_α line and the wavelength shift and toroidal plasma rotation velocity, respectively, relative to the center position in the Ohmic heating phase. Reprinted from Fig. 13 of [26] M. Bitter et al. 'Satellite spectra for heliumlike titanium. II', Phys. Rev. A **32**, 3011 (1985). Copyright (1985) by the American Physical Society

process must be used to determine the true velocity profile [39, 40]. The velocity profile evolution [39] of an ICRF heated H-mode plasma from C-Mod obtained with this spectrometer is shown in Fig. 1.10. As can be seen, during the Ohmic phase before 0.7 s, the core rotation is directed counter-current, becomes weakly co-current following application of ICRF heating after 0.7 s, and is strongly co-current during the H-mode phase from 1.32 to 1.52 s. Similar imaging Johann spectrometers have been installed on KSTAR [41], LHD [42], EAST [43] and W7-X [44], and are being considered for ITER and SPARC.

1.1.2 Active Spectroscopy

The previous section reviewed velocity measurements from Doppler shifts of emission lines populated passively by collisional excitation or radiative recombination from plasma electrons. Another technique is to utilize transitions populated by charge exchange recombination (CXR) with neutral hydrogen isotopes provided by active neutral beam injection (NBI). Advantages of this method are that this can provide a spatially localized measurement (where the neutral beam intersects

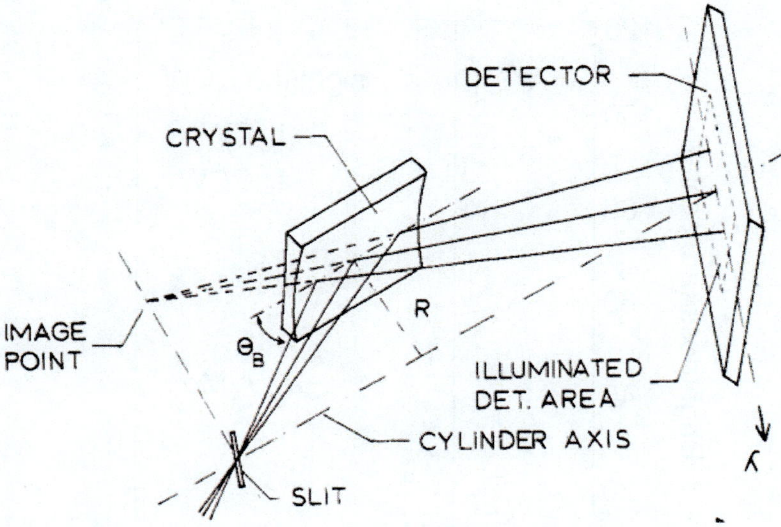

Fig. 1.6 Schematic of the dispersive characteristics of the von Hamos crystal spectrometer used. From Fig. 2 of [32] E. Källne et al. 'High Resolution X-Ray Spectroscopy Diagnostics of High Temperature Plasmas', Phys. Scr. **31**, 551 (1985). ©IOP Publishing. Reproduced with permission.

the spectrometer sightline) and that many transitions populated this way are in the visible wavelength region, which allows the use of visible spectrometers (no vacuum required) and fiber optics. Disadvantages of this technique are that neutral beams constitute a strong momentum source (and perturb the velocity being measured) and that unfolding the line shape to determine the Doppler shift due to the desired plasma rotation can be complicated (for example due to energy dependent atomic cross section distortions [45, 46], see Fig. 1.13). A good review of CXR spectroscopy may be found in [47].

Early observations of CXR population *via* NBI were from H-like oxygen in T-10 [48] and ORMAK [49], and H-like helium in PDX [50]. The first velocity measurements using this technique [51] in the UV spectral range were from ORMAK using the experimental layout shown in Fig. 1.11. The spectrometer system is very similar to that shown in Fig. 1.1; however, there is some spatial localization since CX emission can only occur where the beam crosses the spectrometer line of sight. The observed velocity determined from this instrument for H-like O VIII (O^{7+}) in the UV spectral range due to CXR population from neutral H/D in the n = 1 level agrees very well with the passively measured He-like Ne IX (Ne^{8+}), as can be seen in Fig. 1.12. CXR spectroscopy (CXRS) with NBI (either from heating or diagnostic beams) was soon adapted on tokamaks around the world including Doublet III [52–54], PDX [55], DITE [56], ISX-B [57, 58], ASDEX [59], JET [60] and JIPP TII [61], and is now the standard velocity profile measurement technique. One drawback with this method is that it requires NBI, so the study of intrinsic rotation is tricky.

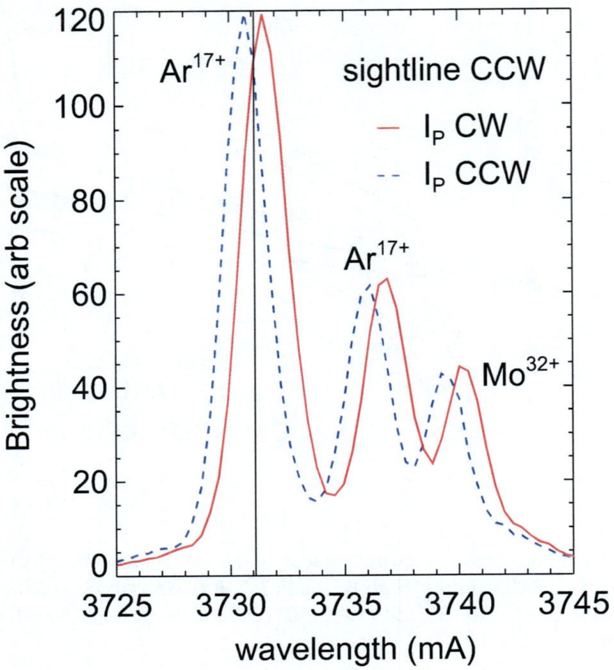

Fig. 1.7 The x-ray spectra of the Ar^{17+} Ly_α doublet and of the $2p^6$-$(2p^5)_{3/2}4d_{5/2}$ transition in Mo^{32+} for counter-clockwise (CCW) I_p (blue dashed curve) and clockwise (CW) I_p (solid red curve) discharges. The $Ly_{\alpha 1}$ wavelength at rest of 3731.1 mÅ is shown by the thin vertical line. Reproduced from Fig. 2 of [34] J.E. Rice et al. 'X-ray observations of central toroidal rotation in ohmic Alcator C-Mod plasmas', Nucl. Fusion **37**, 421 (1997), with the permission of the IAEA

A widely used technique is to utilize beam blips, short duration pulses, to minimize the input torque.

Most applications of CXRS use intrinsic (carbon from walls or boron from wall coating) or injected (*e.g.* neon) impurities which yields the impurity rotation (and temperature). Modern systems utilize CXR from excited neutral hydrogen/deuterium in the n = 2 level, which populate high n (and *l*) levels (for example n = 8 in C^{5+}) which cascade down in the visible spectral region. The impurity rotation can be different from that of background ions, and it is of interest to measure the main ion velocity directly. The first measurements of background ion rotation were obtained in DIII-D helium plasmas [62]. Observations of main ion rotation in D_2 plasmas have also been realized in ASDEX Upgrade [63]. A good review of important issues and complications of CXRS in hydrogen isotopes may be found in [64]. An example of a state of the art CXRS spectrum of D_α is shown in Fig. 1.13 which demonstrates contributions from various effects with and without NBI in a DIII-D H-mode plasma [65]. A comparison of edge rotation profiles determined from carbon and deuterium CXRS using this same system is shown in Fig. 1.14 for a DIII-D H-mode plasma [66]. Clearly there is a substantial difference between

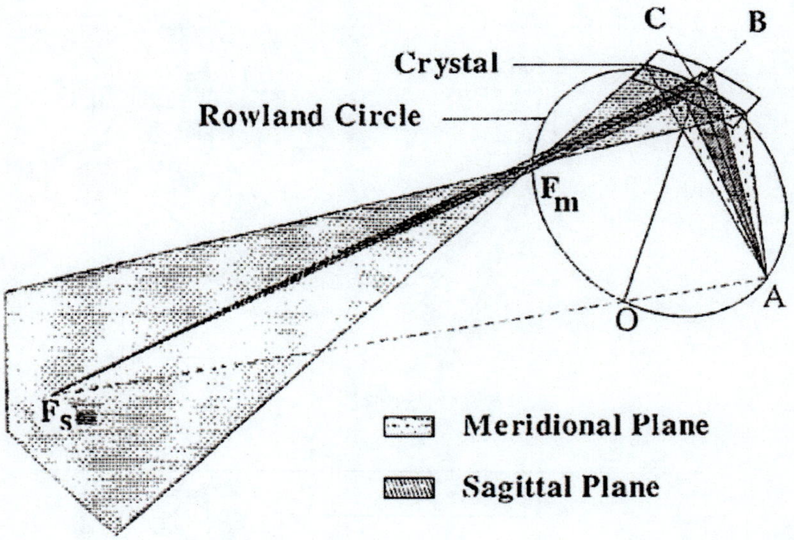

Fig. 1.8 Illustration of the focussing properties of spherically bent crystals. Reproduced from Fig. 1 of [37] M. Bitter et al. 'Numerical studies of the imaging properties of doubly focussing crystals and their application to ITER', Rev. Sci. Instrum. **66**, 530 (1995), with the permission of AIP Publishing

impurity and background ion rotation near the last closed flux surface (LCFS), as can be seen in panel (d). There is an increase in the impurity rotation in the co-current direction outside of the LCFS; this topic will be explored in Chap. 4 on neo-classical effects.

1.2 MHD Mode Rotation

Another method of measuring the velocity is to examine the rotation frequency of n = 1 pre- and post-cursors to sawtooth oscillations using magnetic pickup coils or soft x-ray arrays. This technique was first employed on PLT [16, 67] and has been used subsequently on ISX-B [58], JET [68] and Alcator C-Mod [35, 36, 69–71]. An example of the velocity at the q = 1 surface from sawtooth post-cursors compared to neighboring x-ray Doppler measurements for a C-Mod H-mode discharge is shown in Fig. 1.15 [35, 36]. While the rotation of n = 1 MHD modes relative to ion rotation is subject to interpretation [72], the inferred velocity at the sawtooth inversion radius (r/a $\sim$ 0.2) fits nicely between the impurity rotation at r/a = 0.0 and r/a = 0.3 measured by Doppler shifts of x-ray lines.

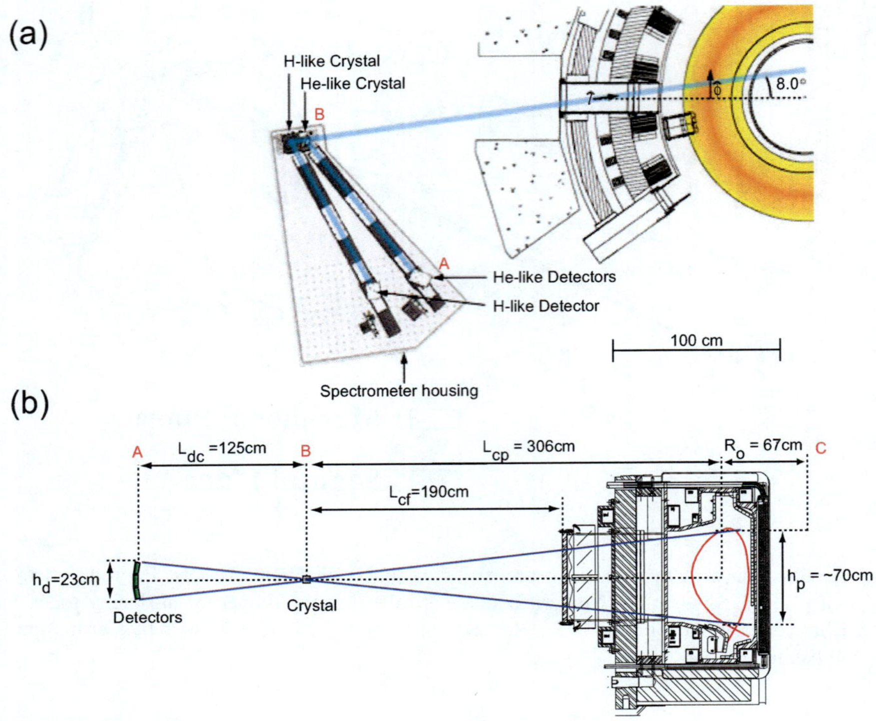

Fig. 1.9 Top view (**a**) and side view (**b**) of the spectrometer as installed on Alcator C-Mod. Reproduced from Fig. 1 of [39] A. Ince-Cushman et al. 'Spatially resolved high resolution x-ray spectroscopy for magnetically confined fusion plasmas', Rev. Sci. Instrum. **79**, 10E302 (2008), with the permission of AIP Publishing

1.3 Probe Measurements

It is possible to utilize Langmuir probe pairs to infer the rotation velocity [73] in the scrape off layer (SOL) of tokamak plasmas [74–77]. The layout of several probe systems on C-Mod is shown in Fig. 1.16 [74]. The EAST and WEST tungsten electrodes, which are embedded in an electrically floating molybdenum body, sample plasma from opposite directions along the same field line, forming a Mach probe pair in which the parallel Mach number can be estimated from the ratio of ion saturation current densities, $M_{\|} = 0.43\, J_{EAST}/J_{WEST}$ [73]. By fitting positive- and negative-going I-V characteristics from the EAST and WEST electrodes, electron densities, temperatures and parallel Mach numbers along the probe's trajectory are obtained every 0.25 ms (corresponding to ~0.25 mm of probe travel). The NORTH and SOUTH tungsten electrodes were operated in a floating voltage mode. The vertical scanning probe is of similar construction and is operated in the same manner. Scanning Langmuir-Mach probe measurements have uncovered clear evidence of strong ballooning-like transport and associated transport driven

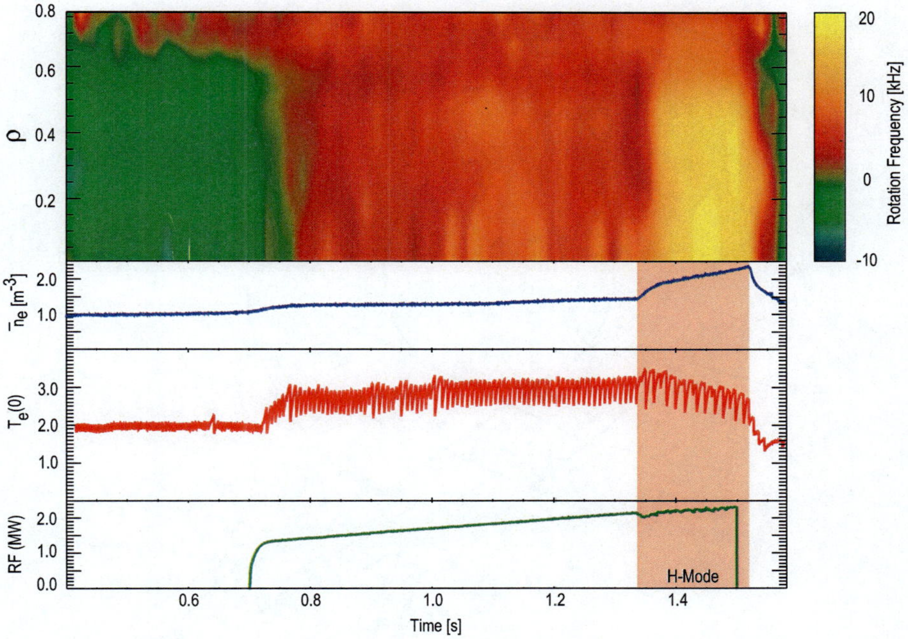

Fig. 1.10 Contour plot showing the rotation profile evolving during an ICRF heated plasma. The line averaged density, central electron temperature and RF heating power are also shown. Reproduced from Fig. 6 of [39] A. Ince-Cushman et al. 'Spatially resolved high resolution x-ray spectroscopy for magnetically confined fusion plasmas', Rev. Sci. Instrum. **79**, 10E302 (2008), with the permission of AIP Publishing

plasma flow along field lines in the inner high field side SOL. A compilation of cross field plasma profiles, including flow measurements, in the inner and outer SOL regions for upper single null (USN), double null (DN) and lower single null (LSN) equilibria is shown in Fig. 1.17 [75]. There is a clear change in sign of the inner SOL flow in going from USN to LSN while the kinetic profiles remain the same. This dependence of rotation on the magnetic equilibrium will be discussed in detail in Sect. 5.2.1.

1.4 Microwave Doppler Reflectometry and Scattering

The Doppler shifts of microwaves may also be used to determine rotation velocities [78–80]. In Doppler reflectometry the antenna tilt angle θ_t induces a Doppler frequency shift $f_D = u_\perp\ 2\ \sin\theta_t/\lambda_0$ in the measured spectrum which is directly proportional to the rotation velocity $u_\perp = v_{ExB} + v_{ph}$ of the turbulence moving in the plasma. Figure 1.18 shows a schematic of the ASDEX Upgrade (AUG) Doppler reflectometer diagnostic along with a poloidal cross section of a typical LSN plasma

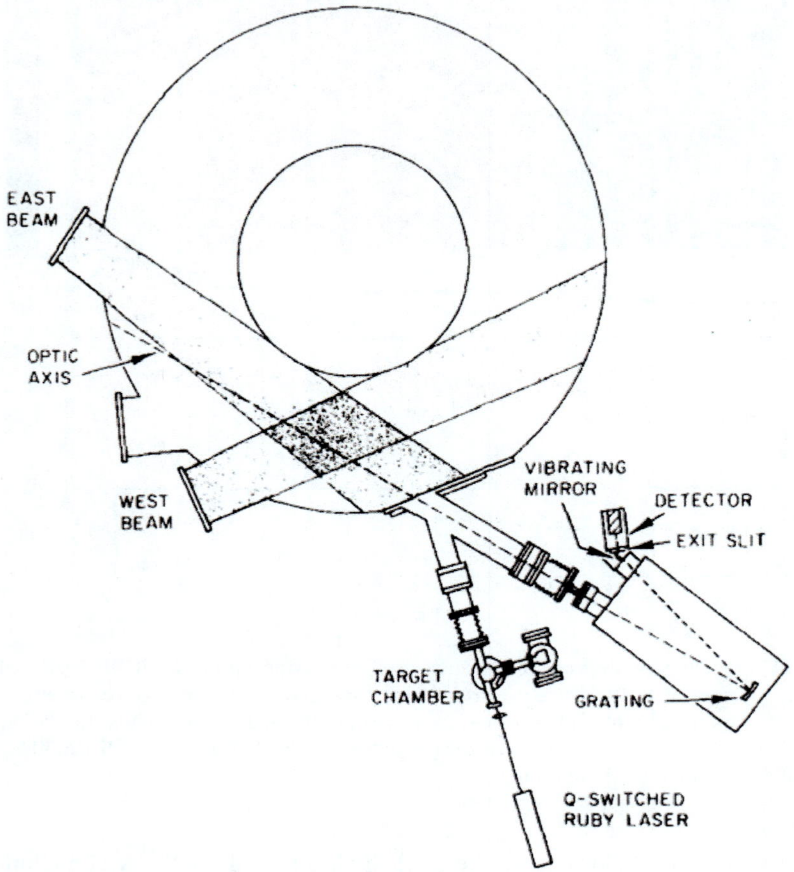

Fig. 1.11 Experimental arrangement for measuring plasma rotation from vacuum ultraviolet lines. The laser ablation apparatus for injecting test impurities is also shown. Reproduced from Fig. 1 of [51] R.C. Isler and L.E. Murray, 'Plasma rotation measurements using spectral lines from charge transfer reactions', Appl. Phys. Lett. **42**, 355 (1984), with the permission of AIP Publishing

[80]. Full details of the diagnostic can be found in [79]. Examples of edge velocity profiles obtained with this system for several AUG Ohmic discharges with various electron densities and plasmas currents are shown in Fig. 1.19 [80]. The profile displays the usual peak structure near the separatrix for all cases, but the interior rotation switches from the electron drift direction to the ion drift direction as the density is increased for the 0.8 MA cases. This rotation reversal will be discussed in detail in Sect. 5.2.3.

Another method of velocity determination relies on collective Thomson scattering (CTS) of microwaves. This technique has been used to measure the background ion temperature [81–83] and may also provide the rotation velocity [84, 85]. An example of the time history of the rotation velocity determined by CTS in an AUG

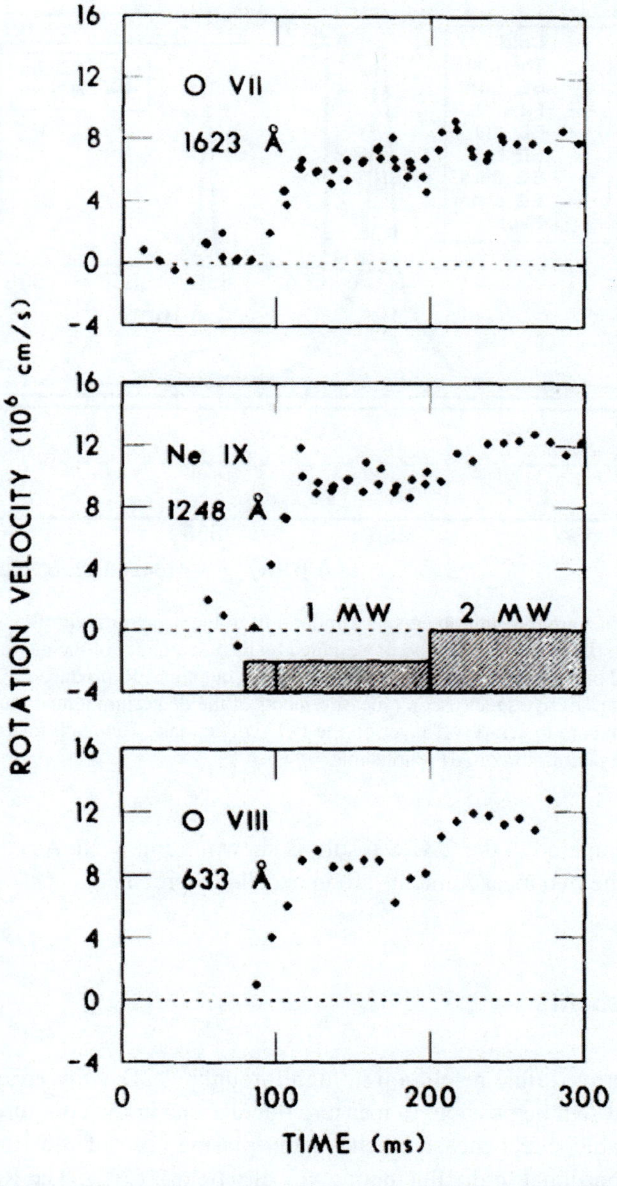

Fig. 1.12 Toroidal rotation velocities measured from charge exchange excited line of O VIII and from electron excited lines of O VII and Ne IX. Reproduced from Fig. 3 of [51] R.C. Isler and L.E. Murray, 'Plasma rotation measurements using spectral lines from charge transfer reactions', Appl. Phys. Lett. **42**, 355 (1984), with the permission of AIP Publishing

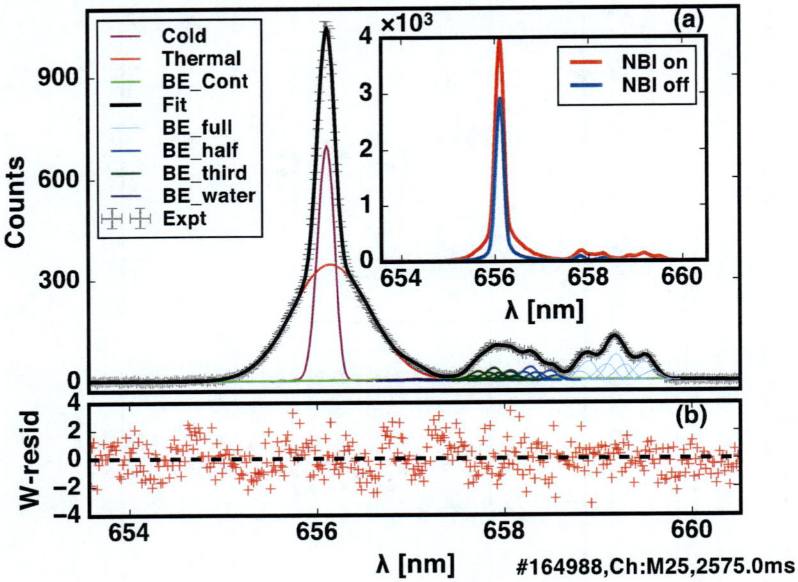

Fig. 1.13 (**a**) Background subtracted spectrum and fit to the spectrum with the NBI on and NBI off spectra inset. (**b**) Residual from the fit weighted by the uncertainty of the measured spectrum at each pixel based on photon statistics, dark noise, and readout noise. Reproduced from Fig. 3 of [65] S.R. Haskey et al. 'Active spectroscopy measurements of the deuterium temperature, rotation, and density from the core to scrape off layer on the DIII-D tokamak', Rev. Sci. Instrum. **89**, 10D110 (2018), with the permission of AIP Publishing

discharge, compared to the CXRS result, is shown in Fig. 1.20. As can be seen, the values from the two measurements are in excellent agreement.

1.5 Comments

With the improved time resolution in high throughput 2D x-ray arrays, soon to be below 1 ms, it will be possible to measure fluctuations in the core toroidal velocity. This will enable direct measurement in the plasma core of the Reynolds stress, which is proportional to the fluctuating velocity fields $\langle \tilde{v}_r \tilde{v}_\phi \rangle$. The Reynolds stress in the plasma boundary region has been observed directly using probes [76, 77].

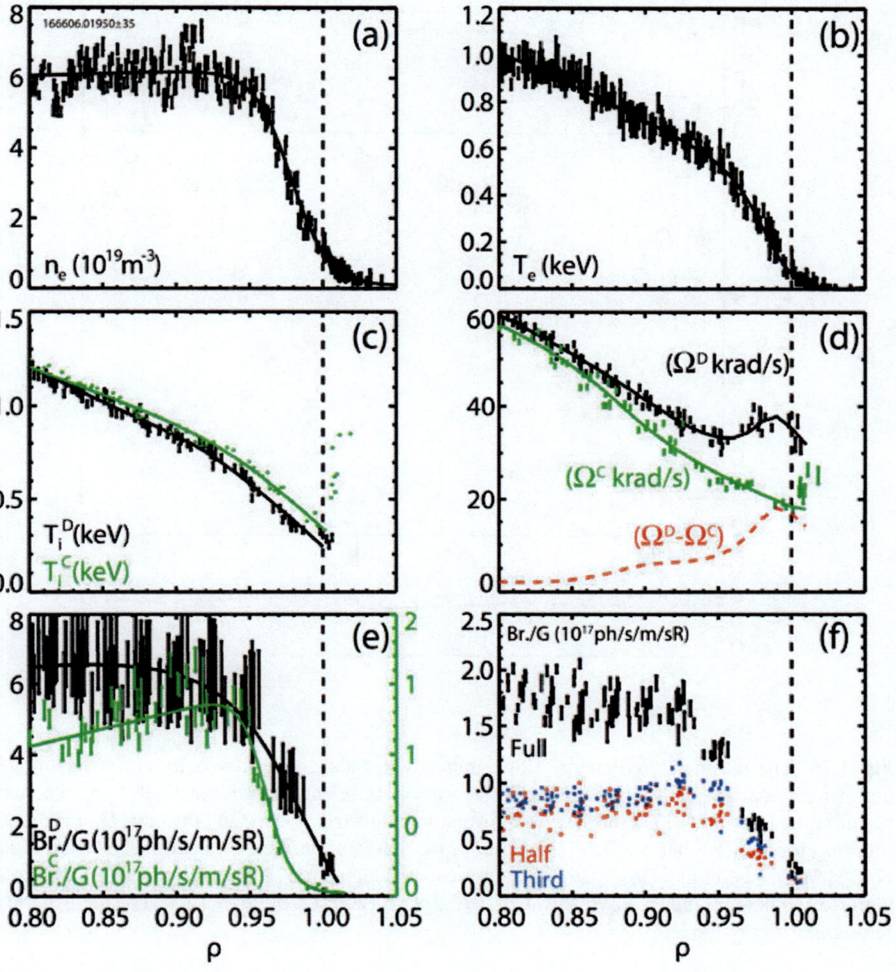

Fig. 1.14 H-mode pedestal profiles from Thomson scattering, impurity, and main ion CXRS. Carbon and deuterium rotation frequency profiles are shown in panel (**d**). Reproduced from Fig. 4 of [66] B.A. Grierson et al. 'High resolution main-ion charge exchange spectroscopy in the DIII-D H-mode pedestal', Rev. Sci. Instrum. **87**, 11E545 (2016), with the permission of AIP Publishing

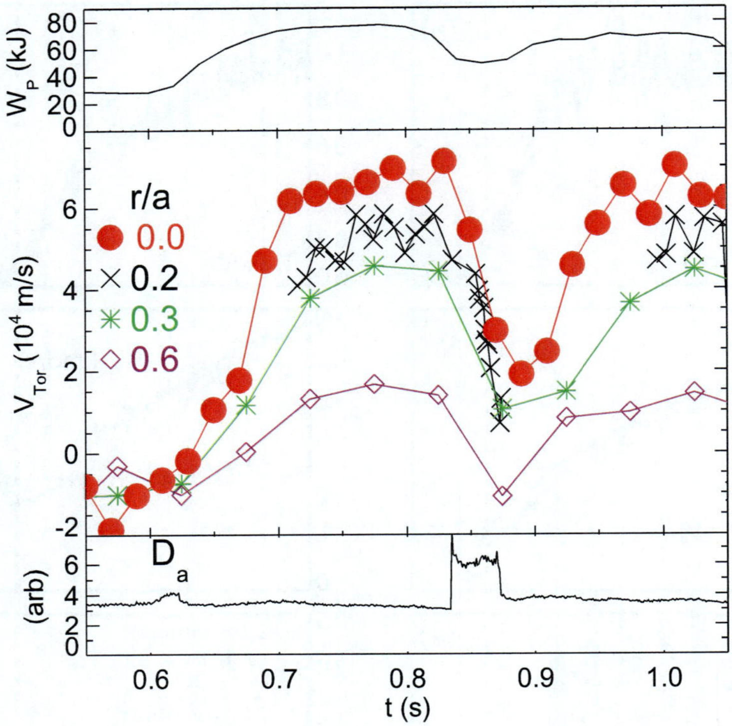

Fig. 1.15 The plasma stored energy (top), impurity toroidal rotation velocity at three radii (red dots, green asterisks and purple diamonds for r/a = 0.0, 0.3 and 0.6, respectively) and magnetic perturbation rotation (×s) at the sawtooth inversion radius (r/a ∼ 0.2), and the edge D_α brightness (bottom) for an ICRF heated ELM-free H-mode discharge. Reprinted with permission from Fig. 2 of [35] W.D. Lee et al. 'Observation of Anomalous Momentum Transport in Tokamak Plasmas with No Momentum Input', Phys. Rev. Lett. **91**, 205003 (2003). Copyright (2003) by the American Physical Society

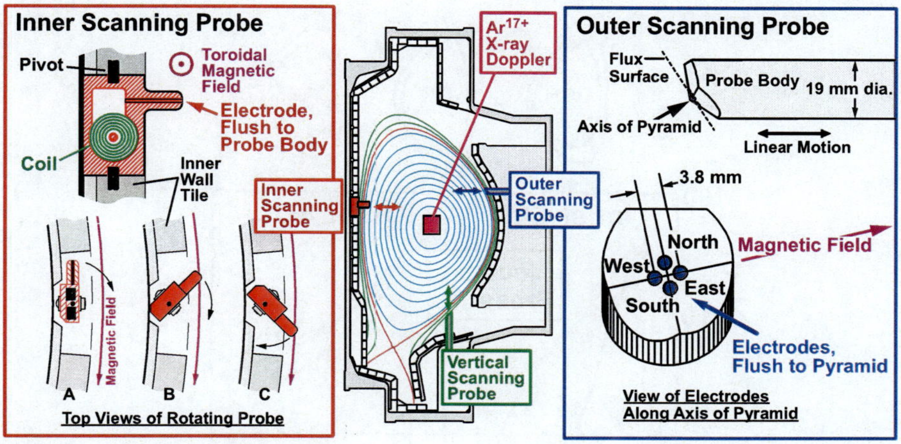

Fig. 1.16 Scanning Langmuir-Mach probes are used to infer simultaneously plasma profiles and parallel flows at three locations in the SOL of Alcator C-Mod, including the high field side (inner scanning probe). A typical lower single null plasma equilibrium is shown. Reproduced from Fig. 1 of [74] B. LaBombard et al. 'Transport-driven Scrape-Off-Layer flows and the boundary conditions imposed at the magnetic separatrix in a tokamak plasma', Nucl. Fusion **44**, 1047 (2004), with the permission of the IAEA

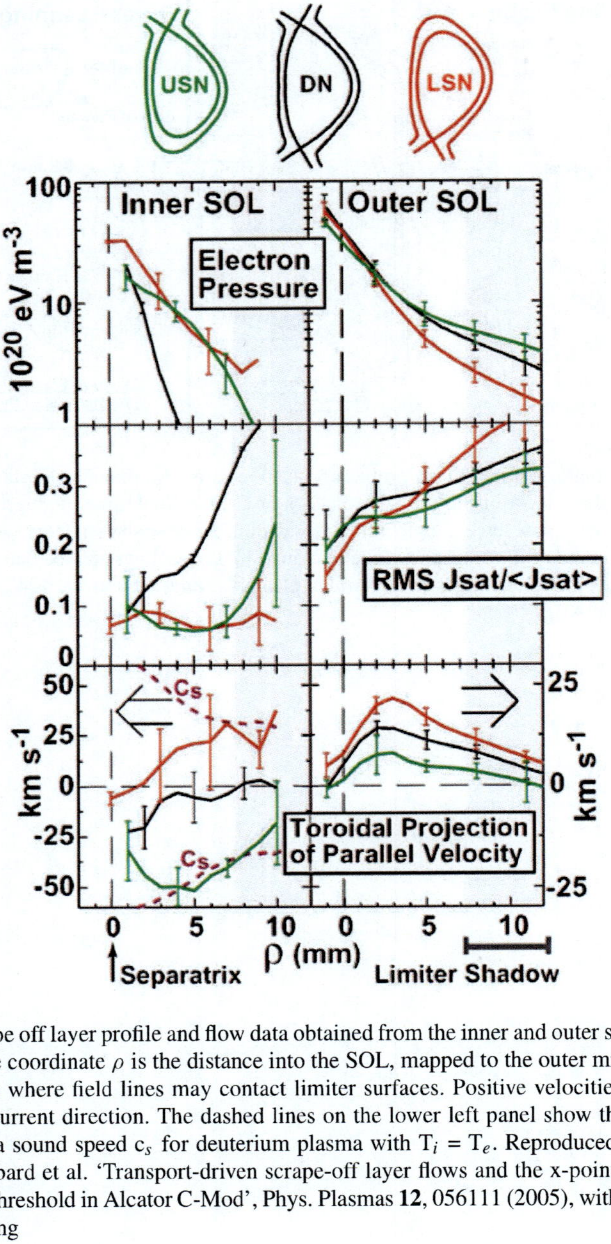

Fig. 1.17 Scrape off layer profile and flow data obtained from the inner and outer scanning probes. The flux surface coordinate ρ is the distance into the SOL, mapped to the outer midplane. Shaded regions indicate where field lines may contact limiter surfaces. Positive velocities correspond to flow in the co-current direction. The dashed lines on the lower left panel show the magnitude of the local plasma sound speed c_s for deuterium plasma with $T_i = T_e$. Reproduced from Fig. 2 of [75] B. LaBombard et al. 'Transport-driven scrape-off layer flows and the x-point dependence of the L-H power threshold in Alcator C-Mod', Phys. Plasmas **12**, 056111 (2005), with the permission of AIP Publishing

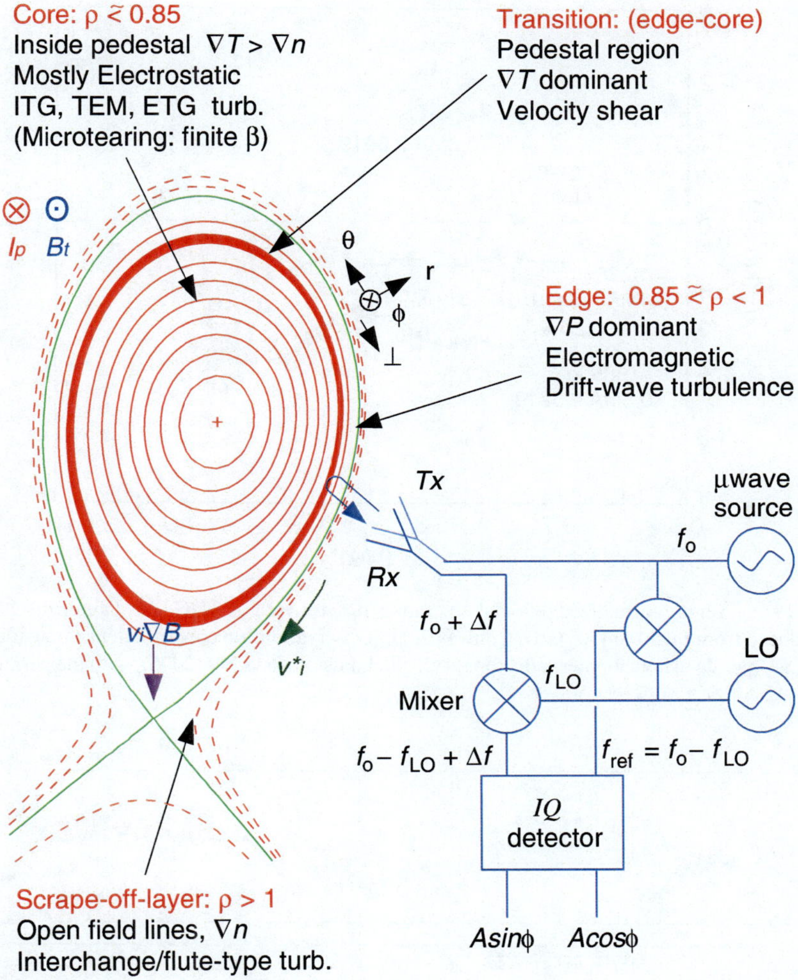

Fig. 1.18 Poloidal cross section of a typical AUG discharge with a schematic of the Doppler reflectometer. Reproduced from Fig. 1 of [80] G.D. Conway et al. 'Observations on core turbulence transitions in ASDEX Upgrade using Doppler reflectometry', Nucl. Fusion **46**, S799 (2006), with the permission of the IAEA

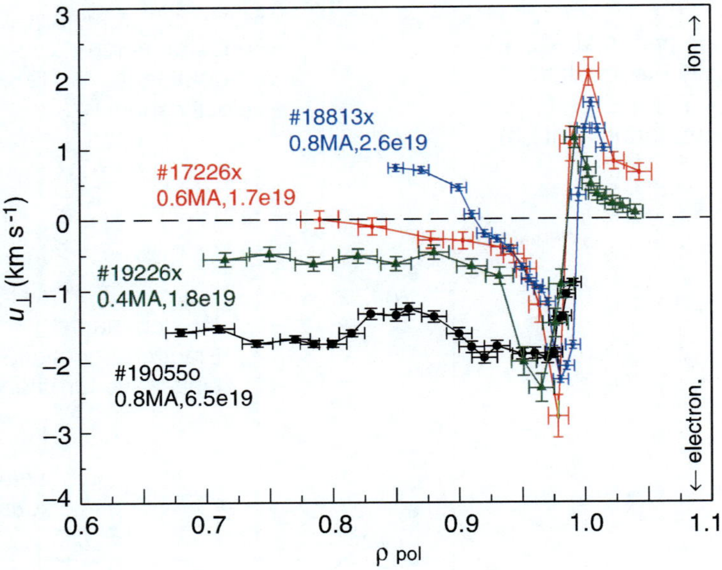

Fig. 1.19 u$_\perp$ versus normalized poloidal flux radius ρ_{pol} for various AUG LSN Ohmic discharges. Reproduced from Fig. 4 of [80] G.D. Conway et al. 'Observations on core turbulence transitions in ASDEX Upgrade using Doppler reflectometry', Nucl. Fusion **46**, S799 (2006), with the permission of the IAEA

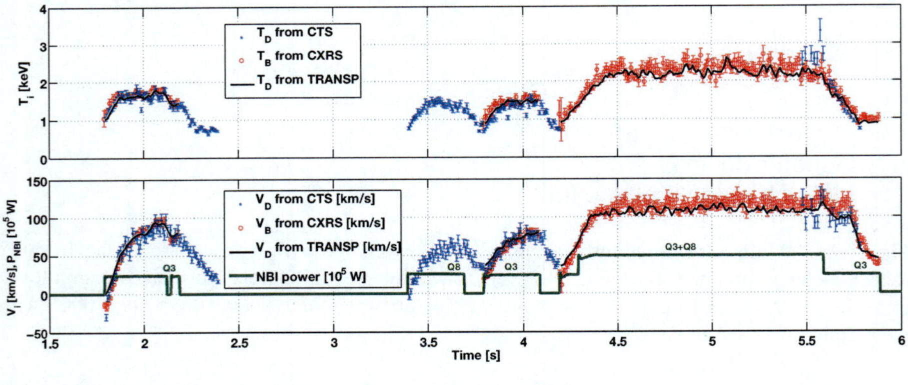

Fig. 1.20 Deuterium ion temperatures (upper panel) and plasma rotation velocities (lower panel) inferred from theoretical fits to the measured CTS spectra and compared to TRANSP results for deuterium and the boron based CXRS measurements mapped to the same flux coordinate. Here positive velocities correspond to co-current (counter B_T) rotation. The error bars indicate plus/minus one standard deviation of the statistical uncertainty in both measurements. The lower panel further shows the total NBI power. Fig. 4 from [84] M. Stejner et al. 'Plasma rotation and ion temperature measurements by collective Thomson scattering at ASDEX Upgrade', Plasma Phys. Control. Fusion **57**, 062001 (2015). ©IOP Publishing. Reproduced with permission. All rights reserved

Chapter 2
Momentum Sources

With the velocity profile determined by use of one or more of the techniques described in the previous chapter, rotation drive may be used to characterize momentum transport, the subject of Chap. 6. In order to achieve this, momentum sources must first be understood. These sources fall into two general categories: direct momentum input from particle or wave interactions (Sect. 2.1) and indirect drive such as from the $\mathbf{j}\times\mathbf{B}$ forces arising from non-ambipolar radial particle shifts or orbit losses (Sect. 2.2), the latter due to toroidal magnetic field ripple for example. Direct rotation drive can be from neutral beam injection, transfer of momentum from radio frequency waves (ICRF, LH or EC) or from neutral gas or compact torus injection.

2.1 Direct Rotation Drive

The most fundamental momentum source is from direct external drive, which can be realized with a variety methods, such as neutral beam injection (NBI) Sect. 2.1.1 or from radio waves of various frequencies: ion cyclotron range of frequencies (ICRF) Sect. 2.1.2, lower hybrid (LH) Sect. 2.1.3 and electron cyclotron (EC) Sect. 2.1.4. Miscellaneous techniques, such as neutral gas injection or compact torus injection (CTI) will be treated in Sect. 2.1.5.

2.1.1 Neutral Beam Injection

The earliest and most studied external rotation drive technique is from utilizing NBI [13, 14, 16, 18, 26, 27, 51, 54, 56, 58, 60, 86–122]. For tangential NBI, the subsequent rotation direction is the same as the NBI [14] orientation, and can

J. Rice, *Driven Rotation, Self-Generated Flow, and Momentum Transport in Tokamak Plasmas*, Springer Series on Atomic, Optical, and Plasma Physics 119,
https://doi.org/10.1007/978-3-030-92266-5_2

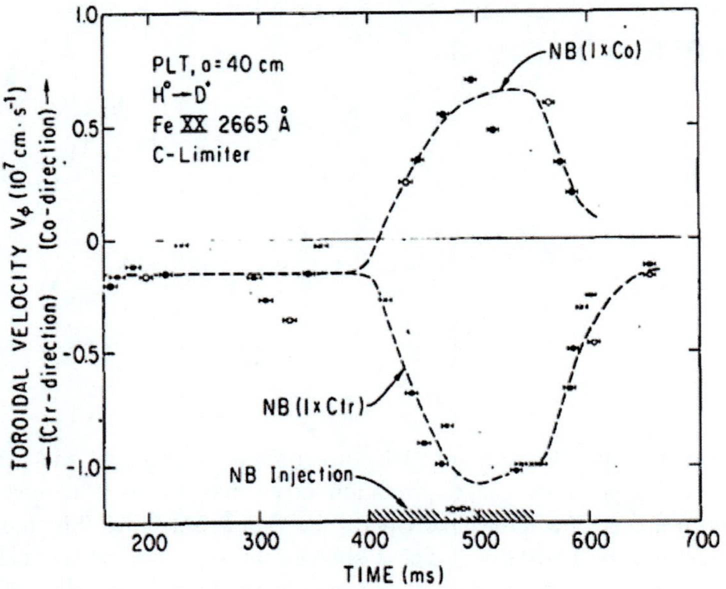

Fig. 2.1 Central toroidal plasma velocities induced by one co-current beam ($P_b \approx 400$ kW) and one counter-current beam ($P_b \approx 500$ kW) from 400 to 550 ms. Reproduced from Fig. 3 of [14] S. Suckewer et al. 'Toroidal plasma rotation in the PLT tokamak with neutral-beam injection', Nucl. Fusion **21**, 1301 (1981), with the permission of the IAEA

be either co- or counter-current, as is demonstrated in Fig. 2.1, for PLT plasmas. Notice that during the Ohmic phase of these discharges before 400 ms, the rotation was directed counter-current; this will be discussed in detail in Sect. 5.2.3. Central co-current velocities up to 700 km/s have been observed in JET plasmas [103] with tangential NBI. In order to model the velocity profile evolution and extract momentum transport coefficients, it is necessary to characterize the input torque from the beams. The dominant direct drive mechanism is usually from the high energy passing NBI ions through collisions with the bulk plasma. It is also important to account for the (nearly instantaneous) indirect collisionless momentum transfer from trapped ions *via* the $\mathbf{j}\times\mathbf{B}$ forces from the radial displacement current which arises from ion motion along banana orbits [13, 14, 88, 92, 96, 114, 119, 123]. This latter indirect drive can be important and will be treated in detail in Sect. 2.2.1. For example in the plasma shown in Fig. 1.2, about 20% of the total input torque arose from the $\mathbf{j}\times\mathbf{B}$ force. Relevant NBI torque processes are included in various modeling codes, such as NUBEAM [124, 125], PENCIL [126], ASCOT [127, 128] and OFMC [129]. An example of the calculated collisional torque density (from NUBEAM) for a 1.5 MA, 3 T JET H-mode plasma [114] is shown in Fig. 2.2. The beam was modulated at 8.33 Hz (top frame) and the response of the rotation frequency (second frame) at mid radius is apparent. This experiment was designed to determine momentum transport coefficients, which will be discussed in detail

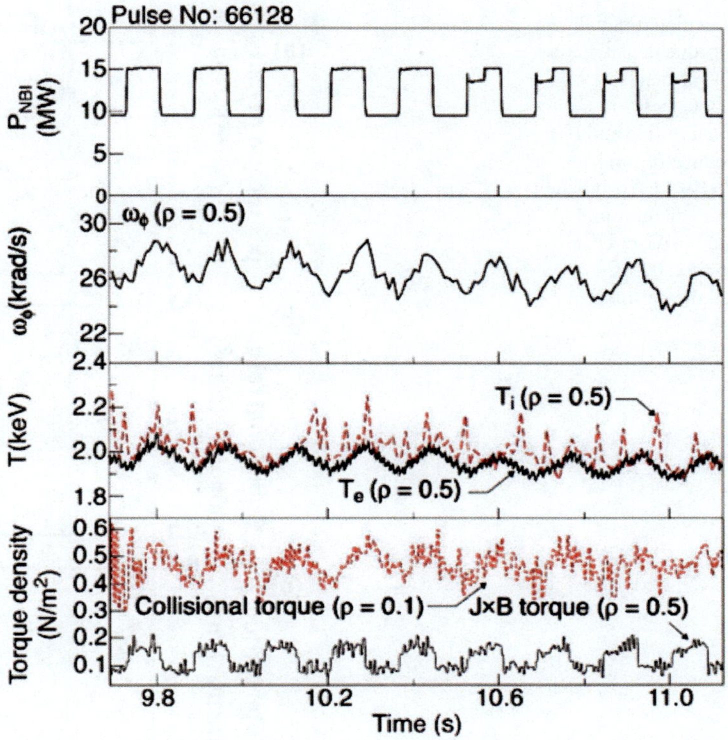

Fig. 2.2 Time traces of the modulated NBI power, rotation frequency, electron and ion temperatures, and two different components of the torque density for JET shot 66128. Reproduced from Fig. 1 of [114] P. Mantica et al. 'Perturbative studies of toroidal momentum transport using neutral beam injection modulation in the Joint European Torus: Experimental results, analysis methodology, and first principles modeling', Phys. Plasmas **17**, 092505 (2010), with the permission of AIP Publishing

in Chap. 6. In the bottom frame is the calculated collisional torque density (at ρ = 0.1). For comparison is shown the calculated **j**×**B** torque density at ρ = 0.5, which will be discussed in Sect. 2.2.1. The collisional torque dominates near the plasma center while the **j**×**B** torque becomes important at mid radius. Another example of the calculated collisional (and **j**×**B**) torque density profiles (from OFMC) for three JT-60U H-mode discharges [119] is shown in Fig. 2.3. These 2.5 T, 1.2 MA plasmas, with densities of 1.8, 2.2 and 3.0 × 10^{19}/m^3, had relatively low direct input torque from balanced NBI. For these three discharges, the collisional torque (B) was directed co-current, while the **j**×**B** torque (A) was counter-current. Since the magnitude of the **j**×**B** torque was comparable to the collisional torque, and because of the opposite scalings with electron density for these two effects, the net input torque (C) was independent of density.

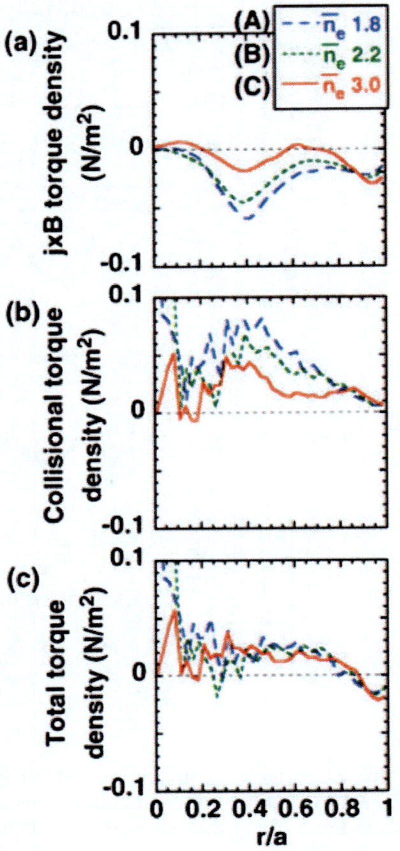

Fig. 2.3 Calculated radial profiles of torque densities for three JT-60U H-mode discharges: (A) $\mathbf{j}\times\mathbf{B}$ torque, (B) collisional torque and (C) total. Reproduced from Fig. 10 of [119] M. Yoshida et al. 'Core and edge toroidal rotation study in JT-60U', Nucl. Fusion **52**, 023024 (2012), with the permission of the IAEA

2.1.2 Ion Cyclotron Range of Frequencies Waves

Radio waves in the ion cyclotron range of frequencies (ICRF, 20–120 MHz) carry momentum and through resonant absorption by fast ions, results in direct collisional and indirect $\mathbf{j}\times\mathbf{B}$ torque on the background plasma. This latter effect, which is related to the NBI induced $\mathbf{j}\times\mathbf{B}$ torque, will be discussed in Sect. 2.2.1. Three types of ICRF waves will be considered: ion Bernstein waves, mode converted ion cyclotron waves and fast magnetosonic waves.

2.1.2.1 Ion Bernstein Waves

Directly launched slow, short wavelength ion Bernstein waves (IBW) have been used to drive poloidal rotation in TFTR D-T plasmas [130, 131]. A comparison of the change in the poloidal velocity profile with and without application of 200 kW of power at 76 MHz is shown in Fig. 2.4 for 3.6 T deuterium discharges from TFTR

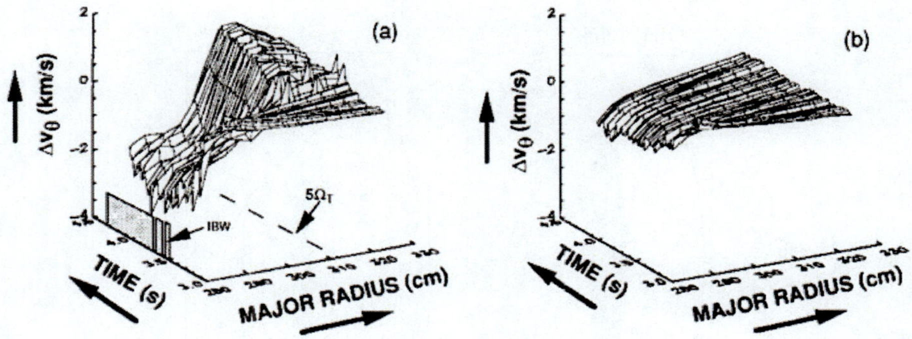

Fig. 2.4 A change in the poloidal rotation velocity as a function of the radius and time. (**a**) IBW heating, P = 200 kW, f = 76 MHz, out of phase excitation, (**b**) no IBW power applied. Reproduced from Fig. 9 of [130] J.R. Wilson et al. 'Ion cyclotron range of frequencies heating and flow generation in deuteriumtritium plasmas', Phys. Plasmas **5**, 1721 (1998), with the permission of AIP Publishing

with a 10% tritium minority [130]. It was observed that the poloidal flow dropped (increased in the electron diamagnetic direction) by ∼2 km/s just outside of the 3rd harmonic deuterium cyclotron resonance layer and increased (in the ion diamagnetic direction) by ∼2 km/s just inside of the 5th harmonic tritium cyclotron resonance layer. Care was taken in these experiments to use ^{6}Li for wall conditioning rather than ^{7}Li, which previously was shown to absorb most of the wave energy due to its charge to mass ratio. The effect of the tritium concentration has also been explored [131]. For lower tritium levels, the change in poloidal flow was negative for all radii between the two harmonic layers. These results were found to be in agreement with a model which treats the Reynolds stress induced by the IBW. From high resolution electromagnetic field calculations which include a compressible Reynolds stress and a second order kinetic pressure model, it has been shown that significant poloidal rotation can be driven by fast magnetosonic or ion Bernstein waves [132]. There are several similar treatments of this mechanism [133, 134].

2.1.2.2 Mode Conversion Flow Drive

In contrast to the case of directly launched IBW discussed in the previous section, it's possible to drive rotation *via* mode conversion of launched fast magnetosonic waves (FW). For example, flow drive has been observed in Phaedrus-T from mode converted kinetic Alfvén waves [135]. Rotation driven by mode converted (MC) ion cyclotron waves (ICW) has been studied extensively on C-Mod [136–138]. To demonstrate evidence for mode conversion flow drive (MCFD), it is informative to compare the rotation results from minority heating (MH) using fast waves and mode conversion into ICW. The setup showing the location of the relevant cyclotron resonance and hybrid layers on the plasma cross section [137] for 5.1 T, C-Mod

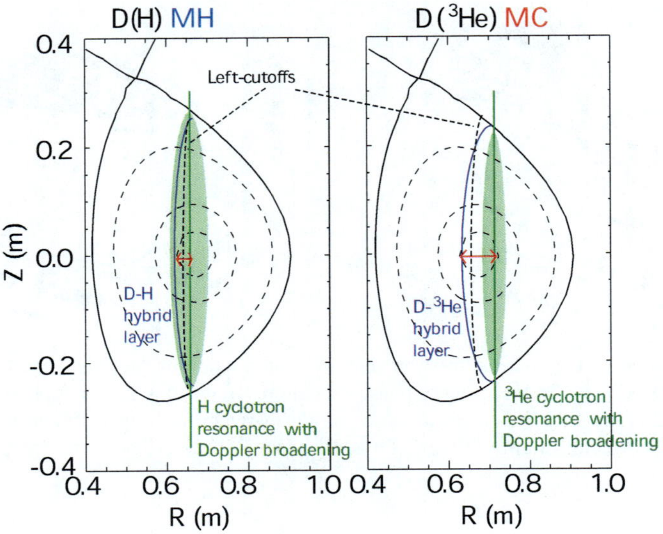

Fig. 2.5 Plasma configuration for the experiment (a) D(H) MH at 80 MHz; (b) D(^{3}He) MC heating at 50 MHz. Both plasmas have $B_T(0) = 5.1$ T, and in upper single null. Reproduced from Fig. 1 of [137] Y. Lin et al. 'Observation of ion cyclotron range of frequencies mode conversion plasma flow drive on Alcator C-Mod', Phys. Plasmas **16**, 056102 (2009), with the permission of AIP Publishing

deuterium discharges is shown in Fig. 2.5. On the left is the MH scheme with a ∼4% hydrogen fraction in a deuterium majority plasma and a launched wave frequency of 80 MHz, and on the right is the MC scheme with a ∼10% ^{3}He fraction and 50 MHz waves. The green regions indicate the Doppler broadened cyclotron resonances and the vertical solid lines are the hybrid layers. The ratio of the distance between these two layers determines whether the scheme is MH or MC. A comparison of poloidal and toroidal rotation time histories under the influence of MH and MC is depicted in Fig. 2.6. There is significant poloidal rotation, nearly 2 km/s in the ion diamagnetic direction, at r/a ∼ 0.4 in the MC plasma which is absent with MH, and the toroidal rotation is nearly a factor of three higher for the same ICRF power, reaching over 80 km/s in the co-current direction [136, 137]. The sheared poloidal flow results from poloidal forces in a narrow layer near the mode conversion surface where the ICW is damped by both electrons and ions, above and below the midplane between r/a ∼ 0.2 and 0.4 [138, 139]. In fact the ICW has been directly detected in these MC plasmas using phase contrast imaging [138]. The mechanism for toroidal flow drive is less clear. In the MH plasma, the toroidal rotation originates at the plasma edge, and is driven by the edge temperature gradient [140] through the residual stress, and will be discussed in Sect. 5.1.2. In the MC plasma, the toroidal rotation appears immediately in the core, and is correlated with the damping of the ICW. Systematic scans of the toroidal magnetic field strength and the ^{3}He concentration have been performed [138] and the magnitude of the velocity is consistent with ICW intensity. Shown in Fig. 2.7 is a comparison of toroidal rotation velocity (top) as a

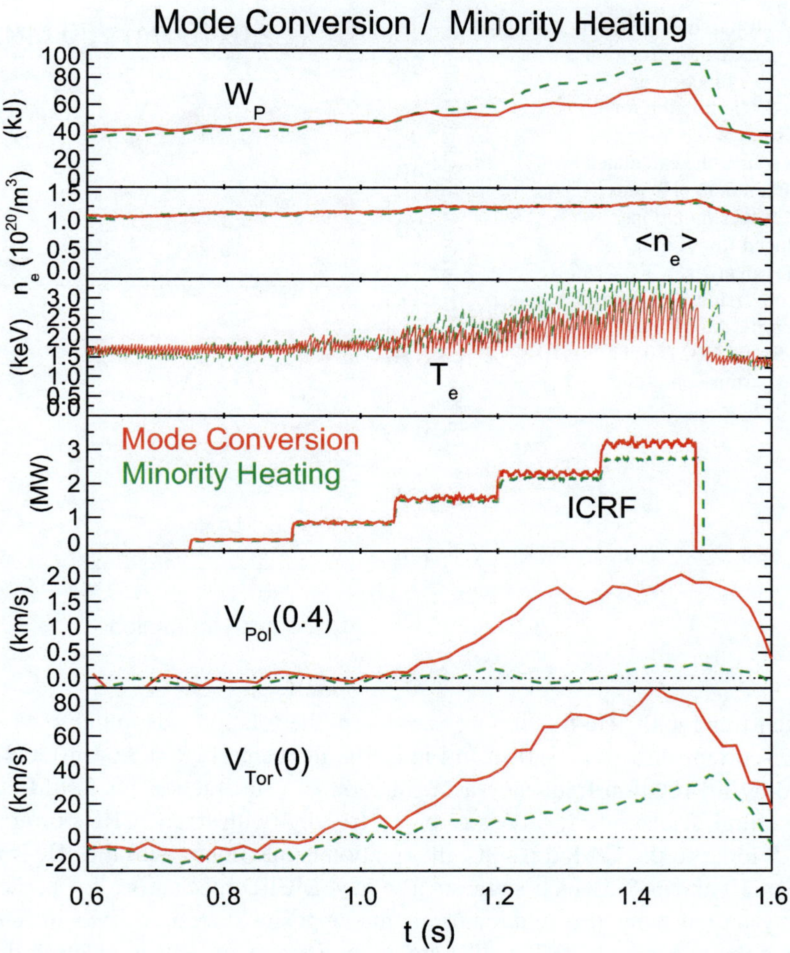

Fig. 2.6 A comparison of ICRF Minority Heating (green dashed) and MCFD (red solid) in 5.1 T discharges from Alcator C-Mod. From top to bottom: stored energy, electron density, electron temperature, ICRF power, poloidal rotation velocity at r/a = 0.4 and central toroidal rotation velocity. Figure 6 from [5] J.E. Rice 'Experimental observations of driven and intrinsic rotation in tokamak plasmas', Plasma Phys. Control. Fusion **58**, 083001 (2016). ©IOP Publishing. Reproduced with permission. All rights reserved

function of the estimated ^{3}He fraction (for 5.1 T, 0.8 MA MC plasmas with n_e = $2.0\times10^{20}/m^3$ and fixed ICRF power at 50 MHz) and the TORIC simulations [138] (bottom) of power to ^{3}He ions from the ICW (same conditions except 1.0 MA). The qualitative agreement is very good, with the observed velocities peaking for ^{3}He fraction between 8 and 15%.

Despite the consistency of this conceptual framework, there are some observations which suggest that the toroidal flow drive mechanism is more complicated. Similar MCFD experiments have been performed in JET plasmas [141], and while

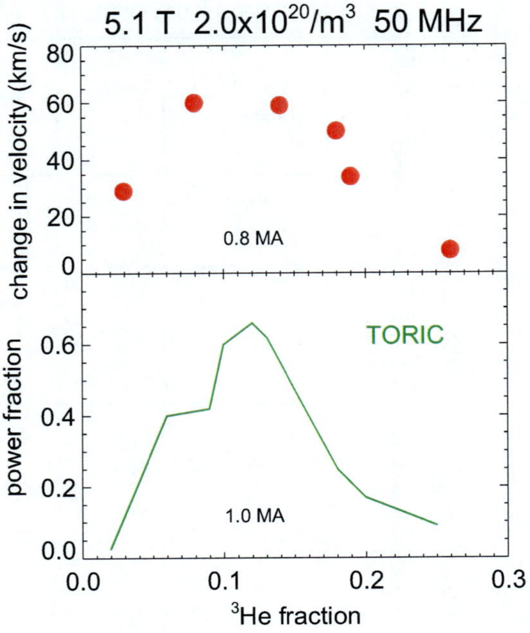

Fig. 2.7 Top frame: observed change in the central rotation velocity as a function of estimated ^{3}He fraction for MC plasmas at 0.8 MA. Bottom frame: the calculated power fraction to ^{3}He ions from TORIC simulations. Reproduced from Fig. 17 of [138] Y. Lin et al. 'ICRF mode conversion flow drive on Alcator C-Mod', Nucl. Fusion **51**, 063002 (2011), with the permission of the IAEA

a similar trend with ^{3}He fraction was observed, the rotation was found to be in the counter-current direction. Shown in Fig. 2.8 is the central (R = 3.04 m) and edge (R = 3.69 m) rotation frequency as a function of ^{3}He fraction for 2.8 MA, 3.45 T deuterium discharges at a density of $3\times10^{19}/m^3$ with fixed ICRF power at 33 MHz. Similar to the C-Mod results, the rotation is maximum with the ^{3}He fraction between 10 and 18%, which suggests a related MCFD mechanism is operational. The mystery is why the rotation is in the opposite direction. Two differences between the C-Mod and JET results are about a factor of ∼10 in electron density and a factor of ∼4 in machine size. There can be no claim in understanding the MCFD mechanism until there is an explanantion of the why the toroidal rotation is in opposite directions in the two cases.

2.1.2.3 Fast Magnetosonic Waves

This section addresses the direct effect of fast magnetosonic waves on rotation, rather than through the intermediate step of mode conversion to ICW or IBW just discussed. RF waves carry momentum, and when absorbed by fast ions, impart toroidal angular momentum $N\hbar$ to them (here N is the toroidal mode number of the wave). The fast ions then transfer the momentum to the bulk plasma by collisions and through a $\mathbf{j}\times\mathbf{B}$ force [142]. This latter effect, similar to the case of NBI, is an indirect momentum source and will be discussed in detail in Sect. 2.2.1. To calculate the collisional torque, several effects must be considered [143]. The propagation and

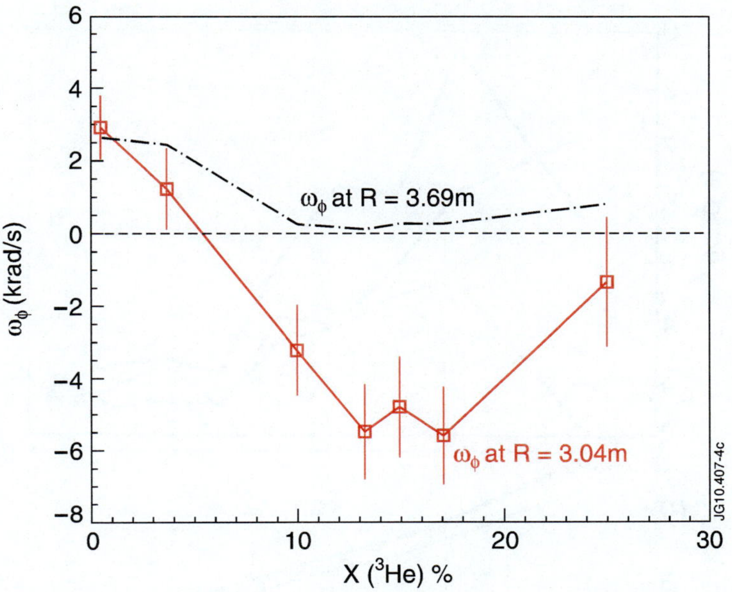

Fig. 2.8 Rotation at R = 3.04 m and R = 3.69 m versus ^{3}He fraction for 3.5 T, 2.8 MA JET discharges. The magnetic axis is at R = 2.97 m and the separatrix is at R ∼ 3.90 m. Figure 4 from [141] Y. Lin et al. 'Ion cyclotron range of frequency mode conversion flow drive in D(3He) plasmas on JET', Plasma Phys. Control. Fusion **54**, 074001 (2012). ©IOP Publishing. Reproduced with permission. All rights reserved

absorption of the ICRF waves need to be calculated and the effect of this absorption on the ion distribution function needs to be included. Since the distribution function affects the dielectric properties required for the wave propagation, a self consistent calculation must be performed. These effects (as well as the indirect $\mathbf{j}\times\mathbf{B}$ force) have been included in the SELFO code [144], which combines a full wave package and a 3D Fokker-Planck solver. There are several other similar treatments of this problem which depend upon the details of the ion orbits through the ICRF wave resonance location [145–153]. A demonstration of plasma rotation induced by directed ICRF waves was performed on JET [154]. Two 3.4 T, 1.8 MA deuterium discharges (with ∼2% ^{3}He) were compared, with 37 MHz waves launched co-current (+90° phasing) and counter-current (−90° phasing), with N ∼ ±12. Rotation and ion pressure profiles are shown in Fig. 2.9. The profile with co-current waves (solid line) has significant co-current rotation compared to the profile with counter-current waves (dashed line), while the ion pressure profiles are quite similar. (The magnetic axis is near R ∼ 3.0 m.) These differences are captured by modeling with the SELFO code, shown in Fig. 2.10. The difference between the two curves of 4 krad/s at r/a ∼ 0.2 is in excellent agreement with the observed difference of 4 krad/s at R = 3.2 m. Bear in mind that there is a net co-current rotation in these plasmas due to different processes (for example residual stress driven by pressure gradients) not included in this modeling. Indeed, other rotation drive mechanisms indirectly induced by ICRF

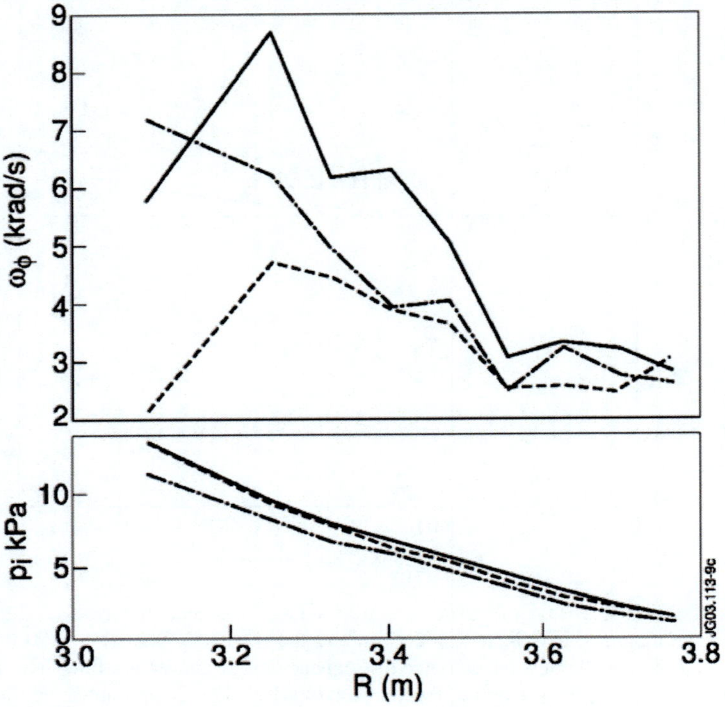

Fig. 2.9 Carbon impurity rotation (top) and ion pressure profiles (bottom) for discharges with +90° (solid line) and −90° phasing (dashed line). (The dot-dashed line is for a discharge with 2 MW of LH power.) Reprinted with permission from Fig. 2 of [154] L.-G. Eriksson et al. 'Plasma Rotation Induced by Directed Waves in the Ion-Cyclotron Range of Frequencies', Phys. Rev. Lett. **92**, 235001 (2004). Copyright (2004) by the American Physical Society

waves can dominate over this direct wave momentum input [71, 155–161], and this will be covered in detail in Sect. 5.1.2. In fact in JET plasmas with off-axis ICRF wave absorption, very little difference in rotation profiles was found with co- and counter-current wave phasing [162].

The toroidal torque provided by the radial electric field due to orbit shifts of high energy ions generated by ICRF waves has been invoked to explain the co-current rotation in ICRF heated H-mode discharges [145, 147, 148, 150]. A particular prediction of some of these theories [147, 150] is that the rotation should switch direction to counter-current with the ICRF resonance located on the high magnetic field side (HFS). The ICRF resonance location has been moved in C-Mod H-mode plasmas by varying the toroidal magnetic field [71, 160, 163]. Shown in Fig. 2.11 is a comparison of two H-mode discharges produced by 1.5 MW of ICRF power at 70 MHz. For the plasma shown in green, the resonance was on the magnetic axis for 4.5 T and there were the usual characteristics of stored energy increase and co-current rotation (Sect. 5.1.2). The discharge shown in red (4.0 T) exhibited very similar co-current rotation even though the resonance location was $(R-R_0)/a$

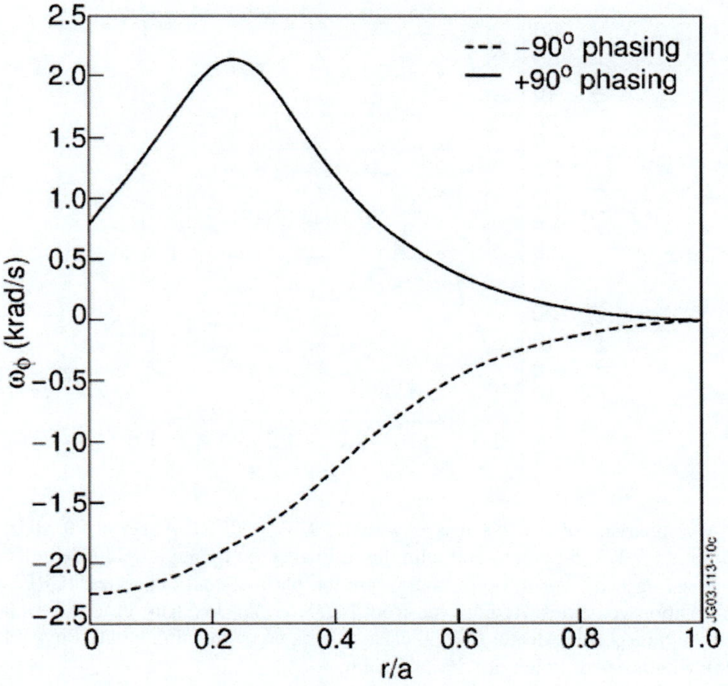

Fig. 2.10 The rotation profiles, ω = V/R, resulting from inserting the fast ion torques simulated by the SELFO code in the momentum diffusion equation. Reprinted with permission from Fig. 4 of [154] L.-G. Eriksson et al. 'Plasma Rotation Induced by Directed Waves in the Ion-Cyclotron Range of Frequencies', Phys. Rev. Lett. **92**, 235001 (2004). Copyright (2004) by the American Physical Society

~ -0.4 on the HFS, in contrast to the predictions. The similarities in the rotation observations (magnitude and scalings) in Ohmic and ICRF heated H-mode plasmas (Sects. 5.1.1 and 5.1.2) suggest that the co-current rotation is not an ICRF wave or fast particle effect.

2.1.3 Lower Hybrid Waves

As demonstrated in the previous Sect. 2.1.2, radio frequency waves carry momentum and can be used to drive rotation. In contrast to the ICRF examples where fast ions absorb the momentum, in the case of lower hybrid (LH, 1–6 GHz) waves, this momentum is absorbed by resonant electrons. The change in the toroidal angular momentum of an electron absorbing energy ΔE from a wave with toroidal wave number n is given by $\Delta P = n\ \Delta E/\omega$. Consequently, the input torque (in the direction the waves are launched) is given by $Rn_{||}P_{LH}/c$ (Nm) [143]. Here R is the device major radius (in m), $n_{||}$ is the parallel index of refraction of the waves, P_{LH} is the LH

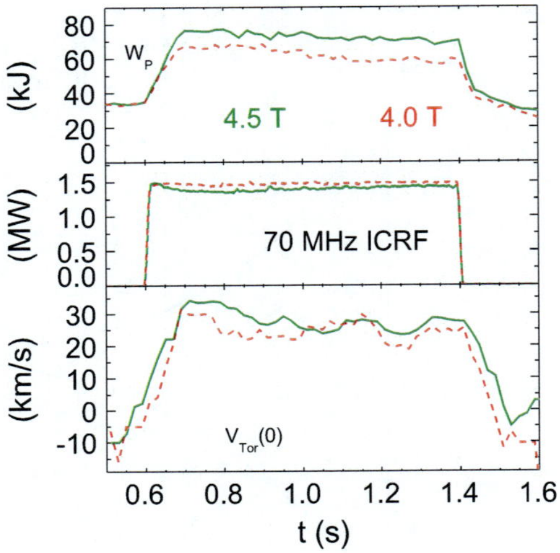

Fig. 2.11 A comparison of two discharges with 1.5 MW of ICRF power at 70 MHz, with on-axis absorption (4.5 T, solid green) and with the resonance on the high field side at $(R-R_0)/a \sim -0.4$ (4.0 T, red dashed). From top to bottom are the plasma stored energies, ICRF waveforms and central rotation velocities. Reproduced from Fig. 5 of [71] J.E. Rice et al. 'Toroidal rotation and momentum transport in Alcator C-Mod plasmas with no momentum input', Phys. Plasmas **11**, 2427 (2004), with the permission of AIP Publishing

power (in W) and c is the velocity of light (in m/s). If the toroidal index of refraction [164] is used, the torque is reduced by about a factor of two. Wave momentum absorption by electrons has been incorporated into the LUKE code [165], which can generate the torque density profiles from LH waves. An example of force density profiles calculated this way is shown in Fig. 2.12 for two 3.8 T Tore Supra (R = 2.34 m, a = 0.78 m) discharges [166] with 4.7 MW of LH current drive (LHCD) power. For the 1.2 MA case, the electron density was $5.9 \times 10^{19}/m^3$ while for the 0.7 MA discharge it was $3.9 \times 10^{19}/m^3$. These force density profiles, which predominantly exhibit rotation in the counter-current direction, are very similar to the LH power deposition profiles.

Counter-current directed rotation [166–173] has been observed in LH experiments on C-Mod, Tore Supra and JET. Direct rotation drive by LH waves is demonstrated in Fig. 2.13, which shows a shot by shot comparison of the time response of the core toroidal rotation velocity to LH wave injection at various power levels for Ohmic L-mode target plasmas in C-Mod [172]. For these cases $n_{||} = 1.9$ and the waves were launched in the counter-current direction, the same as the velocity response. For the discharge shown by the dash-dot line, with 500 kW of LH power, the velocity changed by 35 km/s over 0.5 s (in the inner 1/3 of the plasma, with no change outside of r/a = 0.5). This requires an input torque of order 0.001 Nm, assuming a moment of inertia of the plasma with a flat density profile at $0.7 \times 10^{20}/m^3$ of 1×10^{-7} kg m^2, and assuming a long momentum confinement time. This is comparable to a calculated input torque of 0.001–0.002

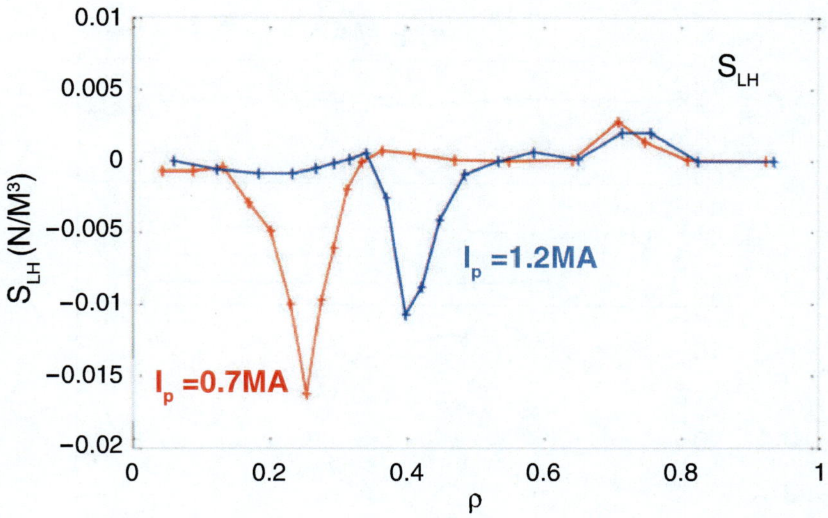

Fig. 2.12 The LH wave momentum transferred to the plasma calculated by LUKE at low (0.7 MA) and high (1.2 MA) plasma current for 3.8 T Tore Supra discharges with 4.7 MW of power. Figure 19 from [166] B. Chouli et al. 'Investigations of LHCD induced plasma rotation in Tore Supra', Plasma Phys. Control. Fusion **57**, 125007 (2015). ©IOP Publishing. Reproduced with permission. All rights reserved

Nm from the LH waves, with R = 0.67 m. Given the direction, magnitude and scaling with power [167, 168, 172], this suggests that the rotation is directly driven by the LH waves, although the details of the time evolution need to be resolved. In spite of this agreement, there are many examples of plasma rotation in the opposite direction to the LH waves [143, 166, 169–175, 177], so the rotation cannot be due only to LH wave induced momentum. This will be treated in Sect. 5.2.2. A number of other mechanisms has been proposed to explain the rotation changes in plasmas with LH wave injection, including: electron orbit loss (see Sect. 2.2.2), the trapped electron pinch [178], resonant electron radial drifts [179, 180], the turbulent equipartition pinch [177] and the effects of magnetic shear on the turbulence spectrum [181–183]. It is likely that there is a simultaneous contribution from many of these effects, and the competition can lead to a rotation reversal. Rotation reversals due to the current density dependence of the residual stress will be discussed in Sect. 5.2.2.

2.1.4 Electron Cyclotron Waves

Similar to IC and LH waves, electron cyclotron (EC, 50–280 GHz) waves carry momentum. In most applications EC waves are used for heating (ECH) and are launched perpendicularly to the toroidal magnetic field, so there is no toroidal momentum input. Just as in the case with ion cyclotron heating, ECH can give rise

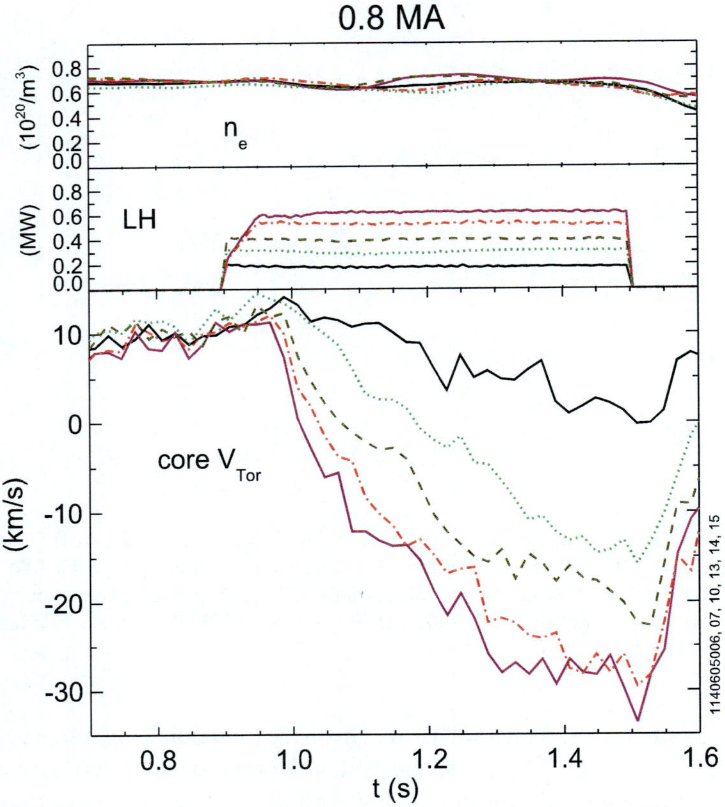

Fig. 2.13 Time histories of the electron density, LH power and core toroidal rotation velocity for 5.4 T, 0.8 MA Alcator C-Mod discharges with different LH power levels. Reproduced from Fig. 1 of [172] J.E. Rice et al. 'Effects of the q profile on toroidal rotation in Alcator C-Mod LHCD plasmas', Nucl. Fusion **56**, 036015 (2016), with the permission of the IAEA

to rotation indirectly, for example through changes in the residual stress. This effect will be treated in Sect. 5.1.3. EC waves can also be used for current drive (ECCD) by launching in the parallel direction and as such can impart toroidal momentum to the plasma. An example of this is shown in Fig. 2.14, a comparison or rotation profiles from DIII-D H-mode plasmas with co- and counter-current ECCD [184]. While the profiles are similar outside of r/a $\sim$ 0.5, the discharge with counter-current ECCD is clearly rotating in the counter-current direction relative to the co-current case in the plasma core. This suggests that while the effects of momentum input from the EC waves are observable, there is another source of momentum in these ECCD plasmas. This is evident based on observations of velocity profile changes under the influence of perpendicularly launched EC heating wave injection.

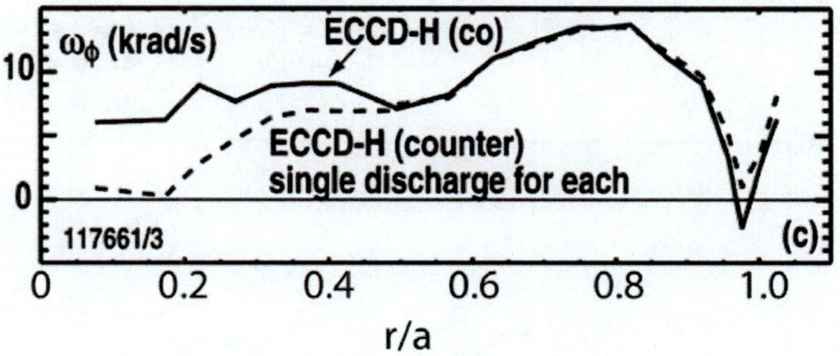

Fig. 2.14 Single examples with co- and counter-current ECCD with core power deposition. Reproduced from Fig. 2 of [184] J.S. deGrassie et al. 'Intrinsic rotation in DIII-D', Phys. Plasmas **14**, 056115 (2007), with the permission of AIP Publishing

2.1.5 Compact Torus Injection

A compact torus can carry momentum [185] and if injected tangentially can drive rotation. Velocity increases of several km/s have been observed in STOR-M L-mode discharges following compact torus injection (CTI) [186, 187] and an example is shown in Fig. 2.15. Following the co-current CTI at 20 ms, the toroidal rotation increased by several km/s at three radial locations. Reversing the CTI relative to the plasma current also reversed the change in rotation [187].

2.2 Indirect Rotation Drive

Indirect rotation drive can arise as a side effect of direct drive through subsequent changes in the radial motion of ion and electron orbits, which gives rise to a radial electric field. $\mathbf{j}\times\mathbf{B}$ forces due to orbit constraints were mentioned in Sects. 2.1.1 and 2.1.2.3, and will be treated in detail in Sect. 2.2.1. Indirect drive due to fast ion and electron loss through expulsion in the toroidal magnetic field ripple will be discussed in Sect. 2.2.2 and the related process from thermal ion loss at the edge will be covered in Sect. 2.2.3.

2.2.1 Orbit Shift $\mathbf{j}\times\mathbf{B}$ Forces

Torque from $\mathbf{j}\times\mathbf{B}$ forces due to radial orbit excursions of trapped energetic ions following capture of beam injected neutrals has been long recognized [13, 14, 16, 88, 92, 123] as an important process in plasmas with low collisionality, ν_* (the ratio between the collision frequency and the bounce frequency). The ions have three constants of motion between collisions: the kinetic energy, the magnetic moment

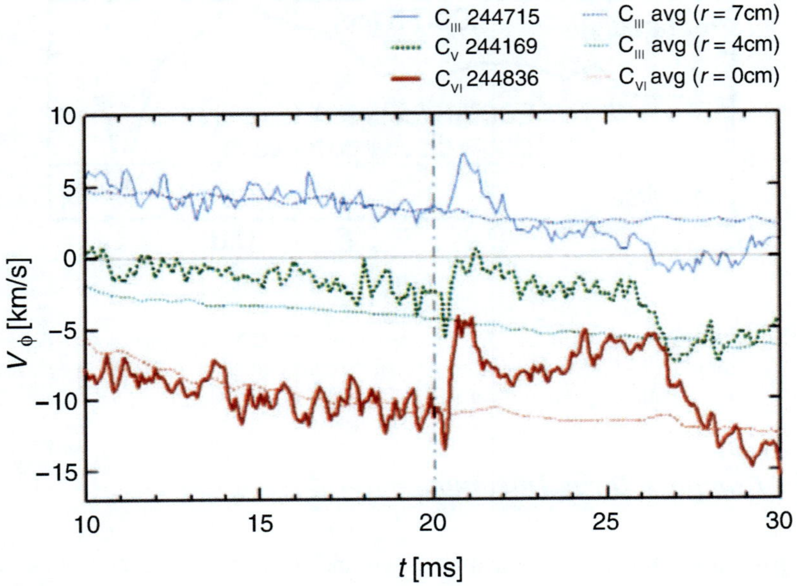

Fig. 2.15 Flow velocities of C III, O V and C VI in the STOR-M tokamak shown by solid (blue), thick dashed (green) and bold (red) lines. The dotted lines show the averaged flow velocity in ten typical Ohmic discharges without CTI. The CT trigger time is t=20.0 ms, indicated by the dashed-dotted vertical line. Figure 6 from [186] T. Onchi et al. 'Effects of compact torus injection on toroidal flow in the STOR-M tokamak', Plasma Phys. Control. Fusion **55**, 035003 (2013). ©IOP Publishing. Reproduced with permission. All rights reserved

and the toroidal component of the canonical angular momentum. The toroidal angular momentum of each ion itself is not a constant of motion, although the toroidal angular momentum is conserved for the plasma as a whole. For ions injected parallel to the current in a tokamak plasma, the drift surfaces are shown in Fig. 2.16 [92]. The toroidal angular momentum, L, averaged over the passing orbit, $\langle L\rangle_p$, is larger than L_I at the inboard point of absorption and smaller than L_O at the outboard point of absorption as $L_I < \langle L\rangle_p < L_O$. The toroidal angular momentum averaged over the trapped orbit, $\langle L\rangle_t$, is much smaller than L_O ($\langle L\rangle_t << L_O$) and is given by the rate of precession of the banana orbit around the torus in the presence of the radial electric field that is set up by the rotating plasma. Since the average minor radius over one drift surface is different from the minor radius at birth, and fast ions are continuously injected, there is a radial current of fast ions. The fast ion radial current is compensated for by an opposite current in the bulk plasma, resulting in a $\mathbf{j}\times\mathbf{B}$ force on the bulk plasma. An outward fast ion current produces a counter current torque on the bulk. Therefore, fast ion loss produces a torque in the counter-direction regardless of the direction of the NBI (co-, counter- or perpendicular). The difference between the toroidal angular momentum at birth and the first orbit average is transfered during one bounce time ($\sim$ 10 μs), which is much faster than the slowing down time. This explains the more rapid increase

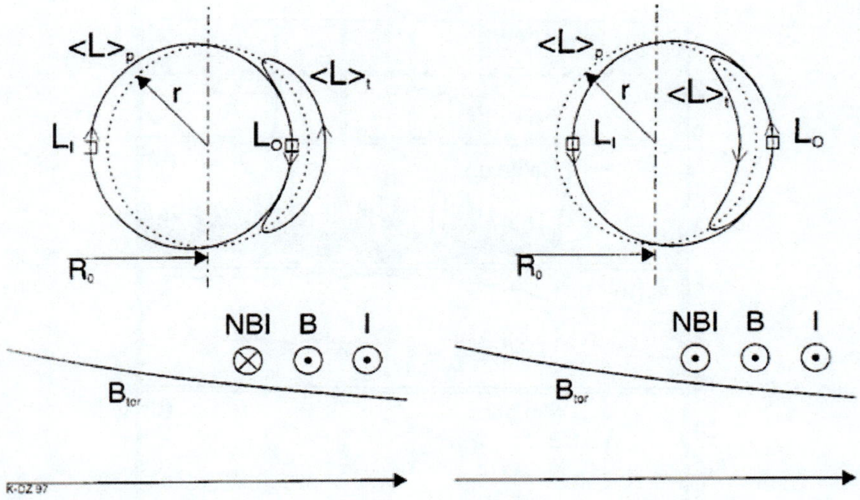

Fig. 2.16 Drift surfaces for passing particles, absorbed inboard or outboard, and trapped particles for counter-injection (left) and co-injection (right). Reproduced from Fig. 4 of [92] K.-D. Zastrow et al. 'Transfer rates of toroidal angular momentum during neutral beam injection', Nucl. Fusion **38**, 257 (1998), with the permission of the IAEA

of the toroidal angular momentum compared to the plasma stored energy following the turn on of NBI power [92]. Isolation of the toroidal rotation driven by the **j**×**B** torque from fast ions has been demonstrated in DIII-D He ECH H-mode plasmas using beam blips which are much shorter than the slowing down time [96]. Time histories for such a discharge are shown in Fig. 2.17. During each blip the velocity is driven up in the direction of the NBI and relaxes back to a lower level by the start of the next blip, while the ion temperature remains unchanged. The integration time for the velocity measurement in the first three blips was 1 ms, short compared to the NBI pulse width of 10 ms, and much shorter than the volume averaged slowing down time (~45 ms) and the D^+ fast ion perpendicular scattering time (~80 ms). Thus, the change in toroidal rotation is due to the **j**×**B** torque and not the collisional transfer of momentum from passing ions during their slowing down process. This **j**×**B** torque is routinely included in neutral beam codes, and examples were shown in Figs. 2.2 and 2.3.

A similar **j**×**B** force can arise due to absorption of ICRF waves, even with a symmetric toroidal wavenumber spectrum. An illustrative example for a particular situation in JET plasmas is depicted in Fig. 2.18 [142]. The resonance interaction takes place where the Doppler shifted wave frequency is an harmonic of the ion cyclotron frequency, ω_{ci}. For a trapped particle, the change in the parallel velocity at the resonance is related to the change in the turning point

$$\Delta \mathrm{v}_\phi = eZB_\mathrm{P}/\mathrm{m}\Delta \mathrm{r}_{\mathrm{t.p.}}. \tag{2.1}$$

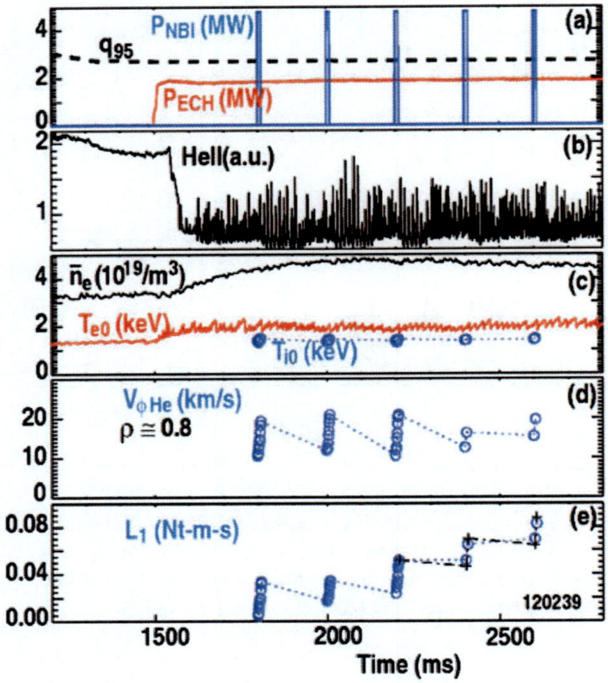

Fig. 2.17 Discharge time traces. (**a**) q_{95}, ECH and NBI power. (**b**) Divertor He II recycling light. (**c**) Line averaged electron density and near central electron and ion temperatures. (**d**) Bulk He^{++} toroidal velocity at $\rho \cong 0.8$. (**e**) Total integrated toroidal mechanical angular momentum, ratcheting up from NBI blip to blip. Reproduced from Fig. 3 of [96] J.S. deGrassie et al. 'Prompt toroidal momentum balance with collisionless neutral beam injected torque in DIII-D', Phys. Plasmas **13**, 112507 (2006), with the permission of AIP Publishing

Because the density of fast resonating ions initially decreases with minor radius, there is an effective diffusive flow of fast ions outward, and the sum of the changes in toroidal velocity is finite, so on average, $\Delta r_{t.p.}$ is positive. This must be balanced by an inward flow of low energy ions. This radial current gives rise to the $\mathbf{j}\times\mathbf{B}$ force which in the case here, is in the direction opposite to the plasma current. Variations of this *scenario* with resonance location, antenna phasing and current direction have been widely studied [146–152] theoretically, and experimental results reveal a complicated situation [162], indicating that several different rotation drive mechanisms are at play. For example, it was predicted that changing the ICRF resonance location from the low to high field side would cause the rotation to switch directions from co- to counter-current [147], which was not observed in experiments [160].

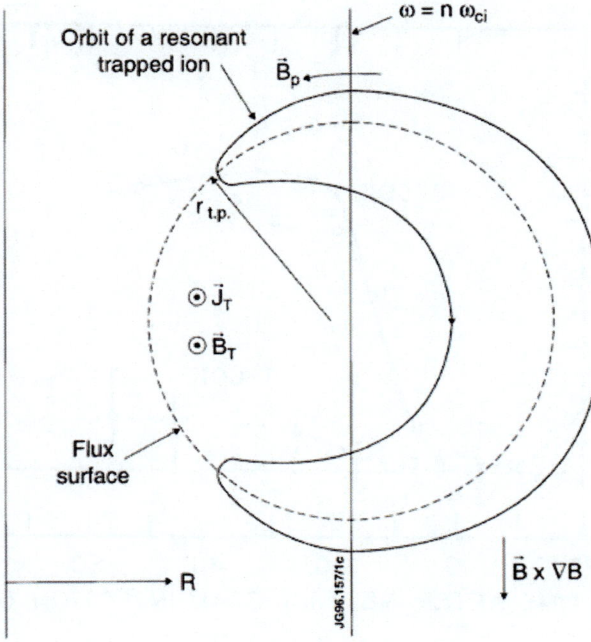

Fig. 2.18 Geometry of IC wave absorption in a torus. The directions of the magnetic field and current are relevant to a particular JET experiment. Figure 9 from [142] L.-G. Eriksson et al. 'Toroidal rotation in ICRF-heated H-modes on JET', Plasma Phys. Control. Fusion **39**, 27 (1997). ©IOP Publishing. Reproduced with permission. All rights reserved

2.2.2 Ion and Electron Loss Due to Toroidal Magnetic Field Ripple

The loss of fast ions in the toroidal magnetic field ripple has been invoked as an explanation for the counter-current increments observed in NBI [61, 174, 188–191] and ICRF [175, 176, 192] heated plasmas. This concept was first demonstrated in ISX-B plasmas [188] through increasing the ripple by energizing only half of the toroidal field (TF) coils. Shown in Fig. 2.19 is a comparison of two NBI heated discharges, one with 9 and the other with 18 TF coils. The plasma with 9 TF coils, and therefore higher magnetic field ripple, had much lower rotation. A systematic scan of the toroidal magnetic field ripple by varying the coil currents revealed a linear relation between rotation and TF ripple amplitude [190] in JET plasmas. Velocity profiles for four different values of the magnetic field ripple [191] for JET H-mode palsmas are shown in Fig. 2.20. As can be seen in panel (d), there is a steady drop in the toroidal rotation across the whole profile as the TF ripple amplitude is increased, while the ion temperature profiles in (c) remain unchanged. The counter-current trend as the ripple increases can be traced to changes in the torque generated by the loss of fast ions from the neutral beam. Calculated torque

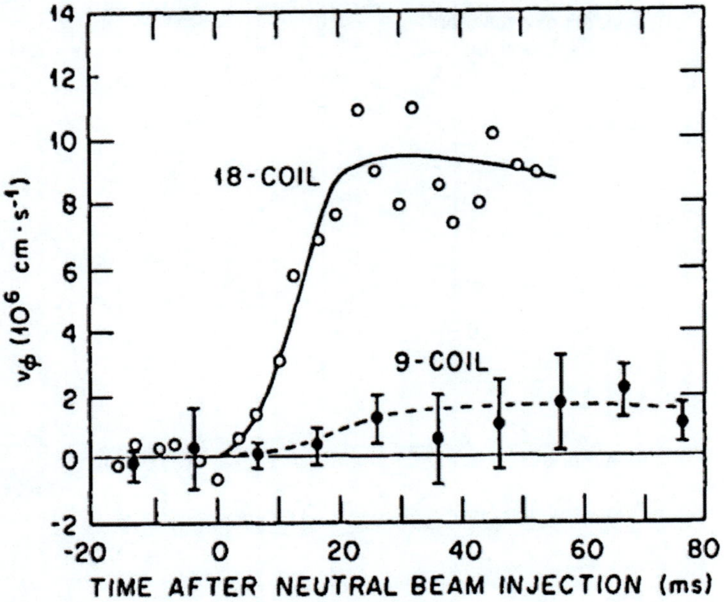

Fig. 2.19 Spectroscopically measured toroidal rotation velocity during neutral injection with 9 and 18 TF coils. Plasma conditions: for 18 coils, $I_p = 130$ kA, $P = 0.95$ MW and $n_e = 2.3\times10^{19}/m^3$; for 9 coils, $I_p = 135$ kA, $P = 0.8$ MW and $n_e = 2.7\times10^{19}/m^3$. Reproduced from Fig. 10 of [188] S.D. Scott et al. 'Measurements of periodic ripple transport in the ISX-B tokamak', Nucl. Fusion **25**, 359 (1985), with the permission of the IAEA

density profiles from ASCOT are shown in panel (a). Most of the changes are ouside of $\rho = 0.7$, and this has a profound affect across the whole profile. (With the ripple loss induced momentum source characterized in this way, varying the magnetic field ripple has been used to determine momentum transport coefficients in JET plasmas [191], which will be discussed in Chap. 6.) Similar results have been obtained in JT-60U plasmas [189] where co-current changes in the edge rotation were correlated with a reduction in the ripple following insertion of ferritic steel tiles, consistent with modeled fast ion loss with the OFMC code. Related observations have been made for the intrinsic rotation in Ohmic Tore Supra [193, 194] and JET [176] discharges, but in these cases, the counter-current change in the rotation was due to thermal ion loss in the TF ripple. For Tore Supra plasmas, the TF ripple was varied by changing the plasma size [194]. Larger ripple leads to more counter-current rotation as ions are lost. Similar behavior has been seen in JET ITB discharges [195]. A corresponding narrative of fast electron loss was used to explain the co-current rotation observed in JT-60U [174] and Tore Supra [175] plasmas with LHCD.

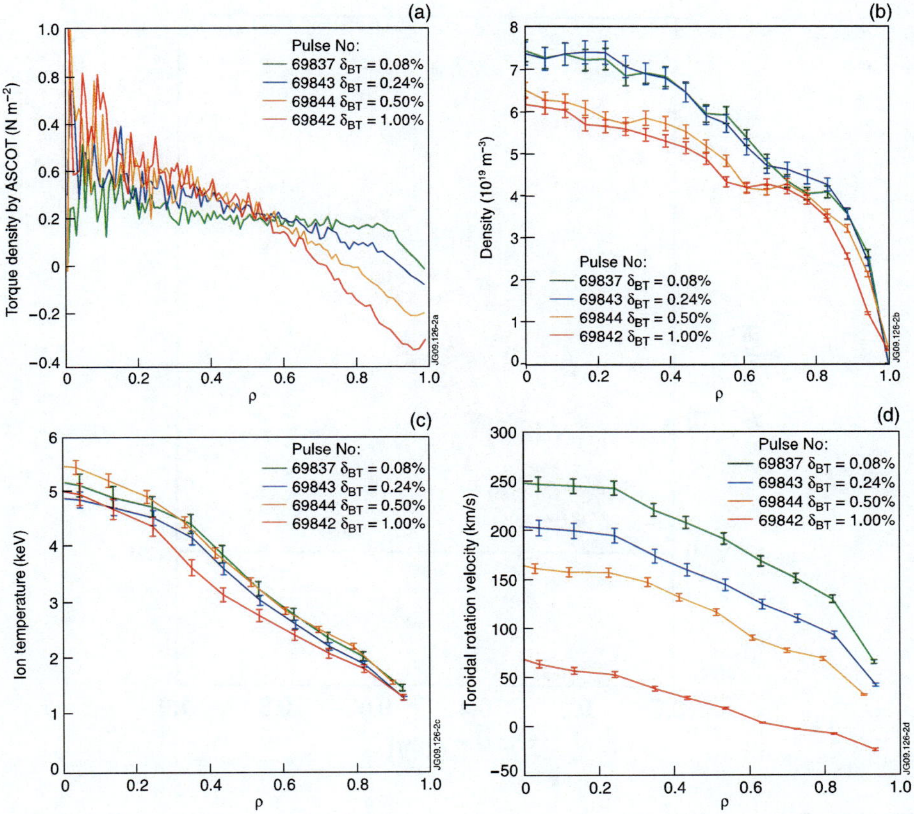

Fig. 2.20 (**a**) The torque deposition profiles as calculated by ASCOT for four similar JET discharges but with different TF ripple amplitudes. (**b**) The corresponding density profiles measured by the Thomson scattering diagnostic. (**c**) The ion temperature profiles measured by CXRS. (**d**) The toroidal rotation profiles neasured by the same diagnostic. Figure 2 from [191] P.C. de Vries et al. 'Momentum transport studies in JET H-mode discharges with an enhanced toroidal field ripple', Plasma Phys. Control. Fusion **52**, 065004 (2010). ©IOP Publishing. Reproduced with permission.

2.2.3 Edge Thermal Ion Orbit Loss

Another ion loss mechanism which gives rise to toroidal rotation is due to the edge thermal ion orbit trajectories themselves. Evidence for this comes from observations of toroidal rotation in the pedestal region of DIII-D H-mode discharges [76, 77, 196–198], where the velocity was found to scale as T_i/B_θ, as can be seen in Fig. 2.21. The magnitude and scaling of the edge velocity is consistent with edge measurements of turbulent stresses [76, 77] and modeling of thermal ion orbit loss [196–207]. Equilibrium flux surfaces and three guiding center loss orbits for these plasmas are shown in Fig. 2.22. Barely trapped counter-current velocity ions starting at the outboard midplane (circles) can be lost, as well as inboard starting co-current

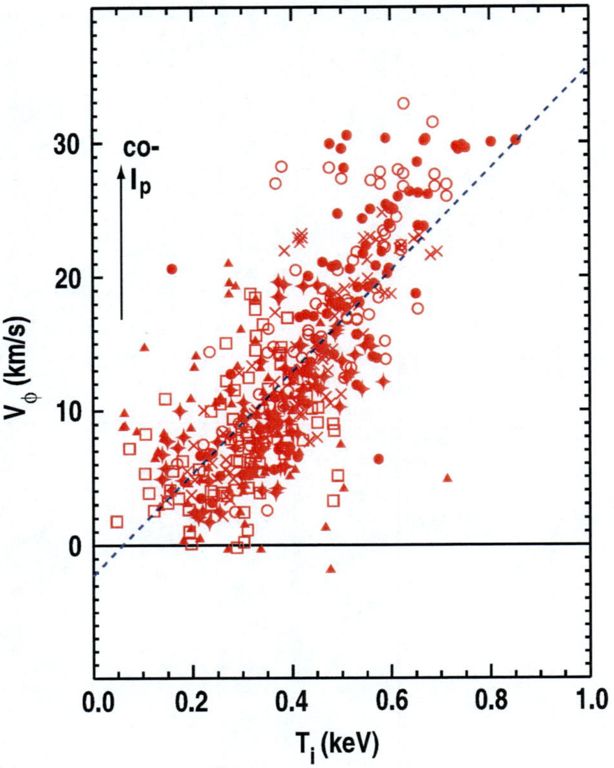

Fig. 2.21 The toroidal rotation velocity as a function of ion temperature in the pedestal region of 51 DIII-D H-mode discharges. Reproduced from Fig. 7 of [196] J.S. deGrassie et al. 'Intrinsic toroidal velocity near the edge of DIII-D H-mode plasmas', Nucl. Fusion **49**, 085020 (2009), with the permission of the IAEA

ions (diamonds) which are in a mirror situation. The more deeply trapped orbits (cusps) have greater energy and thus a greater vertical drift velocity to escape, and can be confined, along with inboard counter-current ions. Simulations with the full f edge code XGC1 reproduce the co-current rotation at the edge [208]. This explanation is strictly valid just inside the last closed flux surface (LCFS). To capture the T_i/B_θ scaling at the top of the pedestal it is necessary to take into account radial displacements of edge passing ion orbits and the interactions with inhomogeneous edge turbulence [209]. Furthermore, a similar T_i/B_θ scaling was derived using higher order neo-classical considerations [210]. Interestingly, both of these results are consistent with the residual stress driven rotation which is found to scale as $\nabla T_i/B_\theta$ (since T_i at the pedestal top is proportional to ∇T_i), and will be discussed in Sect. 5.1.2.

The above discussion was related to edge thermal ion orbit loss giving rise to toroidal rotation. Ion orbit loss can also give rise to poloidal rotation through a viscosity driven flux [211].

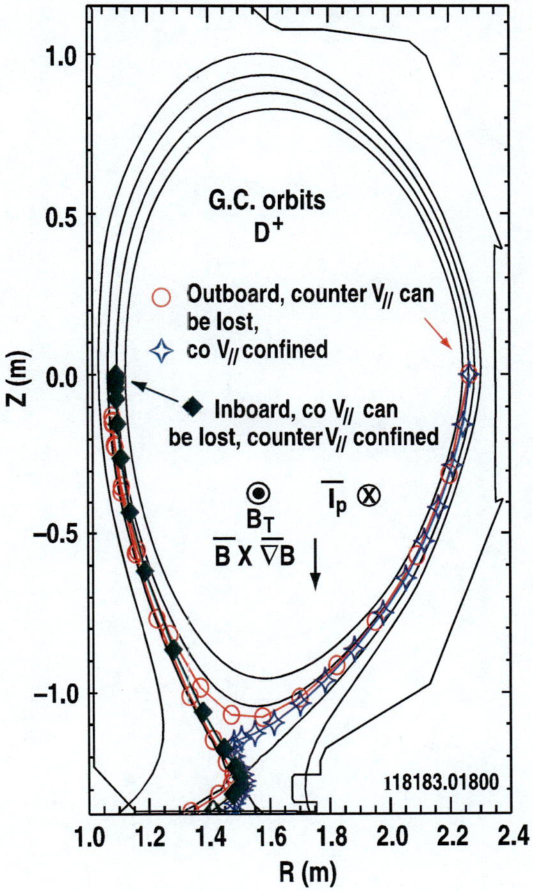

Fig. 2.22 Equilibrium flux surfaces for one of the discharges from Fig. 2.21. Counter-current $V_\parallel$ outboard loss orbits (circles and cusps) and co-current $V_\parallel$ inboard loss orbit (diamonds). These are guiding center orbits computed from the full equilibrium fields. Reproduced from Fig. 8 of [196] J.S. deGrassie et al. 'Intrinsic toroidal velocity near the edge of DIII-D H-mode plasmas', Nucl. Fusion **49**, 085020 (2009), with the permission of the IAEA

2.3 Comments

There is mounting evidence that the observed rotation is not solely due to the direct and indirect drive from various sources, suggesting that intrinsic torque is at play. With balanced co- and counter-current directed NBI (no input torque), there is an offset rotation [91, 100]. Rotation observed in ICRF plasmas is often at odds with the expected momentum input [142, 156]. ICRF mode conversion flow drive in opposite directions has been observed on C-Mod and JET with no known explanation. Similarly, both co- and counter-current rotation is observed in LH plasmas with launched waves only in the counter-current direction. These examples supporting intrinsic torque due to the residual stress will be treated in Chap. 5.

Chapter 3
Momentum Sinks

Working against momentum sources (Chap. 2) is the damping or drag due to various momentum sinks. While momentum sinks are sometimes regarded as sources acting in the direction opposite to the ambient velocity, a true momentum sink is one that damps the magnitude of the rotation regardless of the initial state. Two examples of sinks will be treated in this section: neutral damping (Sect. 3.1) and magnetic braking (Sect. 3.2) due to neo-classical toroidal viscosity (NTV) resulting from non-axisymmetric magnetic fields. The effects of edge localized modes (ELMs) on rotation are considered in Sect. 3.3.

3.1 Neutral Damping

Neutral particles can damp rotation through momentum loss due to collisions with plasma ions. This has been documented in a variety of cases [212–219]. An example of the effect on the rotation by varying the neutral density through different gas puffing rates in JET H-mode plasmas [215] is presented in Fig. 3.1, where momentum density and Mach number profiles are shown. The momentum density and Mach number are seen to drop systematically with increasing gas puffing. Accounting for neutral interactions is complicated due to the inherent 3-D nature of neutral atom distributions. Further complication arises since atomic processes such as ionization and charge exchange recombination must be considered. A schematic representation of the multiple charge exchange interaction process is shown in Fig. 3.2 [215]. When a cold neutral from the edge collides with a plasma ion, it can exchange charge and produce a hotter neutral that scatters unidirectionally. A fraction of these neutrals will penetrate further into the plasma where this process repeats. Subsequently, the thermal energy (a scalar quantity) from the hot neutral is transferred back to the bulk plasma, while the angular momentum of the original ion is conveyed to the background neutrals and lost to the wall. This leads to a difference

J. Rice, *Driven Rotation, Self-Generated Flow, and Momentum Transport in Tokamak Plasmas*, Springer Series on Atomic, Optical, and Plasma Physics 119,
https://doi.org/10.1007/978-3-030-92266-5_3

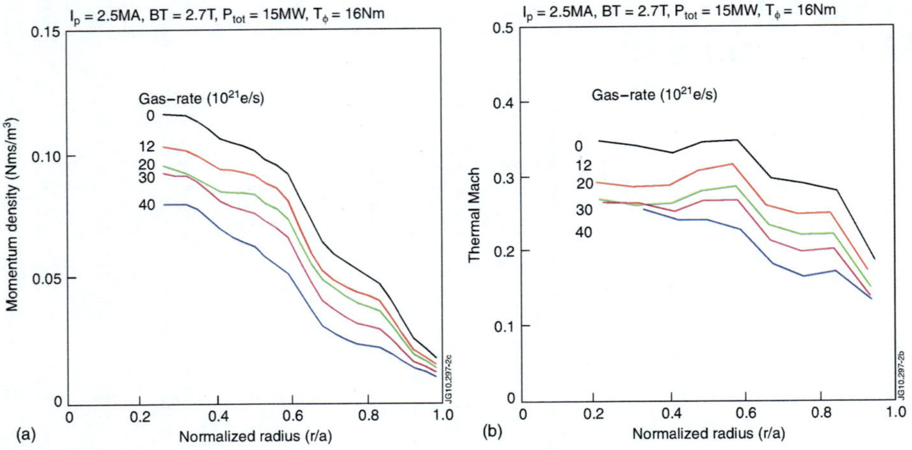

Fig. 3.1 ELM averaged momentum density (**a**) and thermal Mach (**b**) profiles for increasing external gas dosing. A global decrease in momentum density despite an increase in ion density. Notice the large decrease in thermal Mach number in the outer radii. Fig. 2 from [215] T.W. Versloot et al. 'Momentum losses by charge exchange with neutral particles in H-mode discharges at JET', Plasma Phys. Control. Fusion **53**, 065017 (2011). ©IOP Publishing. Reproduced with permission. All rights reserved

in energy and momentum losses through the charge exchange interaction. Taking these processes into account, including modeling of neutral transport, was able to reproduce quantitatively the trends shown in Fig. 3.1. A frictional drag at the plasma edge of up to 10% of the input torque from NBI was deduced, and neutral damping reduced the rotation magnitude for both co- and counter-current beam injection.

3.2 Mode Locking, Magnetic Braking and Neo-Classical Toroidal Viscosity

Breaking of magnetic field axisymmetry can lead to a toroidal viscous drag [36, 60, 104, 105, 109, 110, 122, 220–266]. This damping of plasma rotation from geometric symmetry breaking can be due to a variety of causes: MHD activity, externally applied magnetic fields, toroidal or poloidal field magnet misalignments, magnet bus connexions, or presence of large magnetic objects near the tokamak. An excellent summary of the theoretical background and experimental results can be found in a recent review article [253]. Mode locking of resonant magnetic field perturbations has been shown to stop plasma rotation inside of the q = 2 surface [36, 60, 221, 222, 228]. An example of this is shown in Fig. 3.3 for a JET discharge with a naturally occurring quasi-stationary MHD mode (QSM) [221]. At about 10.5 s there is a large increase in the locked mode amplitude (top frame) and a concomitant damping of the core toroidal rotation velocity (middle frame). A similar

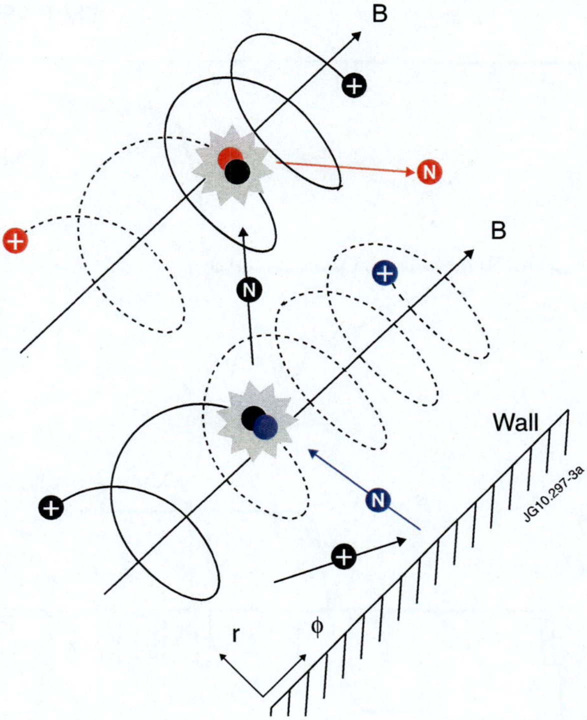

Fig. 3.2 Schematic representation of a multiple charge exchange interaction. Fig. 3 from [215] T.W. Versloot et al. 'Momentum losses by charge exchange with neutral particles in H-mode discharges at JET', Plasma Phys. Control. Fusion **53**, 065017 (2011). ©IOP Publishing. Reproduced with permission. All rights reserved

braking of the rotation velocity can be induced with external magnetic field coils [104, 109, 110, 222, 223, 225, 227–229, 232, 234, 236, 238, 242, 244, 246, 247, 252]. An example of mode locking from energized external coils leading to a stoppage of the core rotation is shown in Fig. 3.4 for two different C-Mod Ohmic L-mode discharges [5]. The target plasmas featured both co- and counter-current intrinsic rotation (due to operation in different Ohmic confinement regimes, which will be covered in detail in Sect. 5.2.3). Following the mode locking at ~0.6 s, the core rotation was reduced to zero in both directions and the sawtooth oscillations were suppressed.

By controlling the amplitude of the resonant and non-resonant magnetic field perturbations *via* external magnetic field coils, it is possible to damp the plasma rotation without reducing it all the way to zero, and without significant losses of particle and energy confinement. Magnetic braking has been investigated extensively in many devices [109, 110, 122, 222, 223, 225, 228–230, 232, 234, 236–240, 242–244, 246–252, 256–260, 267, 268]. A neo-classical 'offset' [110, 231, 233, 234, 236, 240] rotation in the counter-current direction, driven by static non-resonant fields, has

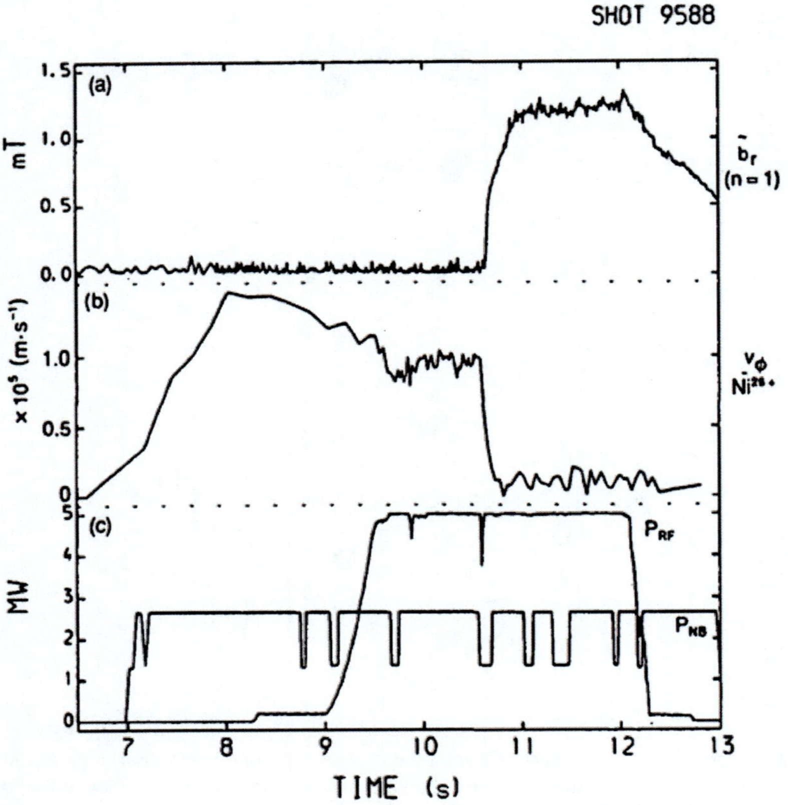

Fig. 3.3 Central ion toroidal rotation velocity from Ni XXVII emission drops abruptly when a QSM locks during combined ICRF and NB heating. (**a**) Locked mode amplitude, (**b**) Ni XXVII ion toroidal rotation velocity, and (**c**) ICRF and NB heating power. Reproduced from Fig. 9 of [221] J.A. Snipes et al. 'Large amplitude quasi-stationary MHD modes in JET', Nucl. Fusion **28**, 1085 (1988), with the permission of the IAEA

been observed in DIII-D plasmas. This is demonstrated in Fig. 3.5 which shows the effect of application of a large amplitude static n = 3 field to a series of target plasmas with different rotation values [234]. When the n = 3 field is applied, braking is observed for initially co- and strongly counter-current rotating plasmas, while an acceleration is seen for the initially weakly counter-rotating discharges. The relation between the measured local torque densities and the toroidal angular velocity after application of the n = 3 field is applied clearly demonstrates that the torque density is proportional to the velocity difference from the offset rotation as $dL/dt \propto (v_\phi - v_{offset})$, where v_{offset} is the rotation due to the neo-classical toroidal viscosity (NTV) caused by the magnetic braking from the n = 3 perturbation field. The observed magnitude, direction and radial profile of the offset rotation are consistent with neo-classical theory predictions.

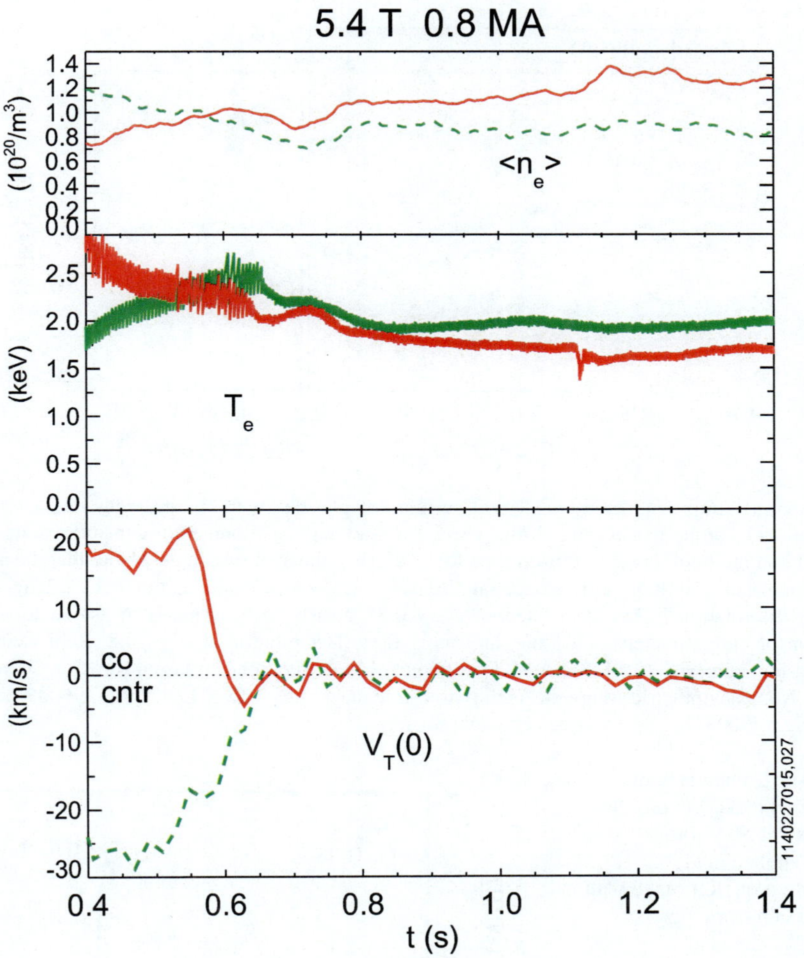

Fig. 3.4 The average electron density (top), central electron temperature (middle) and core toroidal rotation velocity (bottom) for two 5.4 T, 0.8 MA Alcator C-Mod Ohmic L-mode discharges with induced locked modes. Fig. 10 from [5] J.E. Rice 'Experimental observations of driven and intrinsic rotation in tokamak plasmas', Plasma Phys. Control. Fusion **58**, 083001 (2016). ©IOP Publishing. Reproduced with permission. All rights reserved

The theory of NTV is explained in great detail in Ref.[253] and the torque due to NTV can be summarized as $T_{NTV} \propto n_i^{K1} T_i^{K2}$. Values for K1 and K2 depend upon the collisionality regime: K1 = 0, K2 = 5/2 for the $1/\nu$ regime, K1 = 3/2 and K2 = 1/4 for the $\nu-$ sqrt(ν) regime and K1 = 1 and K2 = 0 for the superbanana plateau regime. A quantitative study of NTV has been performed in NSTX neutral beam heated plasmas [228]. The change in the measured angular momentum profile is compared to the theoretical NTV torque by evaluating the angular equation of motion $d(I\Omega_\phi)/dt = T_j$ where T_j is the sum of all torques due to NTV, momentum input from NBI, electromagnetic forces on rotating magnetic islands and viscous

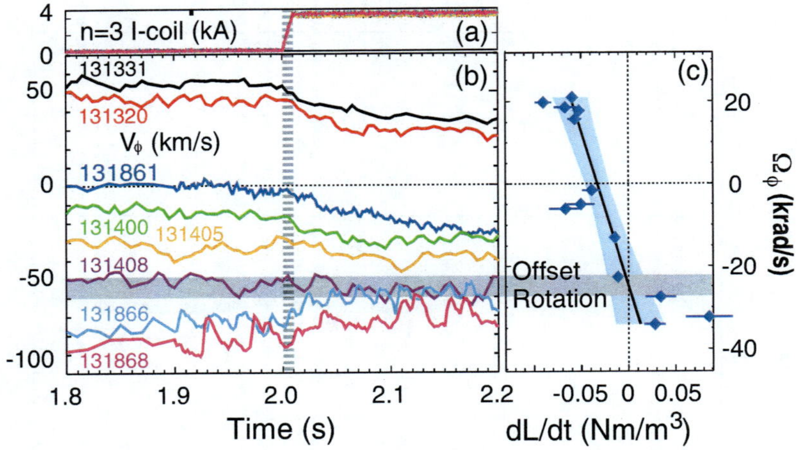

Fig. 3.5 (**a**) Amplitude of n = 3 I coil current for plasmas with (**b**) different initial toroidal rotation. (**c**) Torque density (= dL/dt, where L is the angular momentum density) evaluated at $\rho = 0.8$ at the I coil current turn-on time for a larger number of discharges. Note that the y axis in (**b**) is rotation in km/s at fixed spatial CER view ($\rho \sim 0.8$, major radius R $\sim$ 2.2 m), while in (**c**) it is rotation in krad/s at fixed ρ ($\rho = 0.8$). Nearly zero torque is measured for initial rotation of approximately -50 km/s, indicating the offset rotation at $\rho \sim 0.8$. Reprinted with permission from Fig. 1 of [234] A.M. Garofalo et al. 'Observation of Plasma Rotation Driven by Static Nonaxisymmetric Magnetic Fields in a Tokamak', Phys. Rev. Lett. **101**, 196005 (2008). Copyright (2008) by the American Physical Society

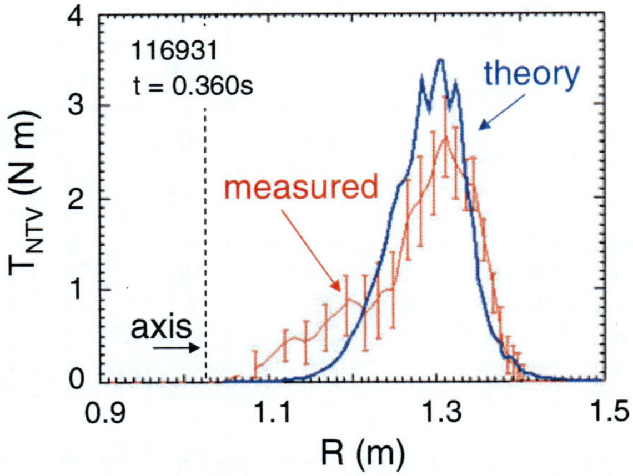

Fig. 3.6 Comparison of measured d($I\Omega_\phi$)/dt profile to theoretical NTV torque for an n = 3 applied field configuration. Reprinted with permission from Fig. 3 of [228] W. Zhu et al. 'Observation of Plasma Toroidal-Momentum Dissipation by Neoclassical Toroidal Viscosity', Phys. Rev. Lett. **96**, 225002 (2006). Copyright (2006) by the American Physical Society

fluid forces between adjacent flux surfaces. In the NSTX experiment, the NTV torque was dominant and is in excellent agreement with calculations, as can be seen in Fig. 3.6. The NTV peaks at R = 1.3 m ($\rho = 0.6$) with a magnitude of $\sim$3 Nm. The magnitude, evaluated from the damping rate of the toroidal rotation, is consistent with the NTV model. Quantitative agreement is found when the effects of toroidally trapped particles is taken into account.

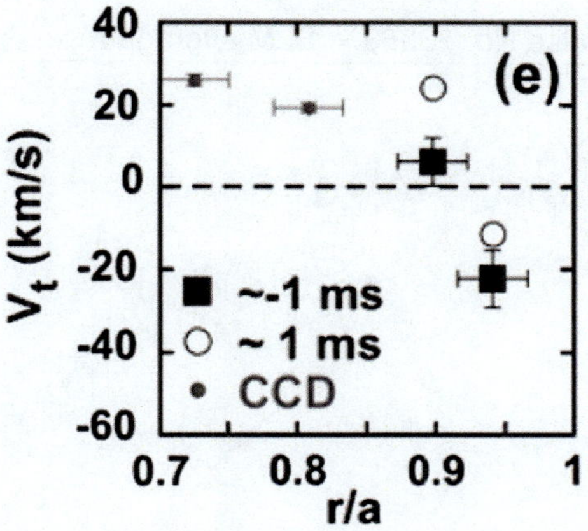

Fig. 3.7 The toroidal rotation profile in the pedestal region of a JT-60U H-mode plasma 1 ms before the ELM (full squares) and 1 ms after the ELM (open circles). Fig. 4 from [269] M. Yoshida et al. 'Observations of the collapse and recovery of the temperature pedestal using diagnostics with the fast temporal resolution in JT-60U', Plasma Phys. Control. Fusion **48**, A209 (2006). ©IOP Publishing. Reproduced with permission. All rights reserved

3.3 Edge Localized Modes

Edge localized modes (ELMs) in H-mode discharges are found to affect the edge toroidal rotation [269, 270]. In the periphery of JT-60U plasmas, a co-current change in the toroidal rotation was observed, as can be seen in Fig. 3.7 [269]. In contrast, at the edge of JET H-mode plasmas, the toroidal rotation profile exhibited a damping in the counter-current direction following ELMs, as can be seen in Fig. 3.8 [270]. Given that the results from the two devices seem to indicate opposite behavior, it is not clear if ELM activity can be regarded as a true momentum sink.

3.4 Comments

Neutral drag has been demonstrated as an important effect opposing external momentum input. Observed rotation damping behavior is generally in excellent agreement with NTV modeling. On the other hand it is uncertain whether ELM activity leads to rotation damping.

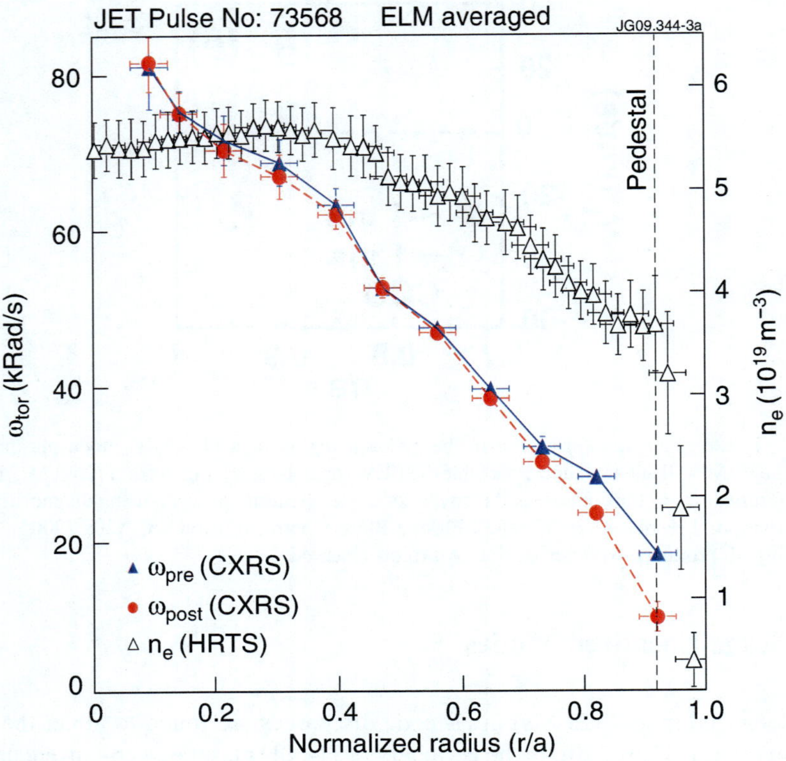

Fig. 3.8 ELM averaged (over 82 individual ELMs) radial profiles and statistical uncertainty pre- and post-ELM for the angular frequency. The pedestal location is shown as derived from the average high resolution Thomson scattering density profile. The error bars denote the statistical spread in the averaging of all profiles within the selected time window. Fig. 3 from [270] T.W. Versloot et al. 'Effect of ELMs on rotation and momentum confinement in H-mode discharges in JET', Plasma Phys. Control. Fusion **52**, 045014 (2010). ©IOP Publishing. Reproduced with permission. All rights reserved

Chapter 4
Comparison with Neo-Classical Theory

In the two previous chapters momentum sources and sinks have been documented. In this section measured velocities under the influence of these sources and sinks will be compared to the predictions of neo-classical theory, which accounts for collisions in the tokamak geometry. The momentum diffusivity (Chap. 6) observed in experiments is generally found to be anomalously large compared to neo-classical levels so it is prudent to check measured toroidal (4.2) and poloidal (4.1) rotation against neo-classical calculations. Expressions for neo-classical main ion and impurity rotation are given in [271–275] and have been incorporated into a variety of codes including NCLASS [276], HAGIS [277, 278], NEOART [279] and NEO [280]. In Sect. 4.3, edge poloidal asymmetries will be addressed.

4.1 Poloidal Rotation v_θ

The impurity poloidal rotation v_θ has been measured in many experiments [12, 17, 20, 21, 62, 98, 195, 281–294] and has been compared to neo-classical calculations, where the dominant dependence is on the ion temperature gradient. The main ion poloidal rotation has a relatively simple expression [271–273],

$$v_\theta = c/eB\ K_1 \partial T_i/\partial r \tag{4.1}$$

where the coefficient K_1 depends upon the collisionality. The calculated K_1 as a function of collisionality for a single impurity species plasma is shown in Fig. 4.1 [273]. Notice that K_1 changes sign for a collisionality of around 10. K_1 also depends upon the relative impurity density, characterized by the impurity strength parameter $\alpha \equiv n_I\ Z_I^2/n_i\ Z_i^2$. Observed poloidal rotation is generally in the electron diamagnetic direction, in qualitative agreement with neo-classical theory, and the

J. Rice, *Driven Rotation, Self-Generated Flow, and Momentum Transport in Tokamak Plasmas*, Springer Series on Atomic, Optical, and Plasma Physics 119,
https://doi.org/10.1007/978-3-030-92266-5_4

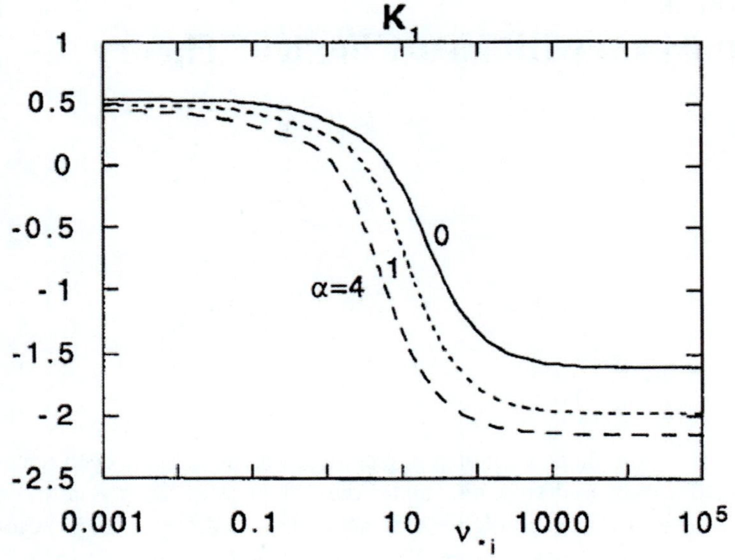

Fig. 4.1 The coefficient K_1 as a function of collisionality ν_{*i} for three values of the impurity strength parameter $\alpha = 0$ (solid), 1 (dotted) and 4 (dashed). Reproduced from Fig. 1 of [273] Y.B. Kim *et al.* 'Neoclassical poloidal and toroidal rotation in tokamaks', Phys. Fluids B **3**, 2050 (1991), with the permission of AIP Publishing

magnitude is usually within a factor of two of predictions. This is considered reasonable agreement by some authors and disagreement by others. An example of excellent agreement between observations, an analytic calculation and several code results is shown in Fig. 4.2 for edge rotation from an ASDEX Upgrade H-mode plasma [292]. At the same time, disagreement between observed and calculated neo-classical v_θ has been reported for Ohmic L-mode discharges on ASDEX-Upgrade [295]. The largest discrepencies between neo-classical calculations and measured poloidal flows are found in plasmas with internal transport barriers (ITBs) [98, 195, 281, 282, 293] where the signs can be opposite, an example of which is shown in Fig. 4.3 [282]. This disagreement suggests that other processes, such as those due to turbulence, can be important in driving poloidal flow [296, 297].

4.2 Toroidal Rotation v_ϕ

Ion toroidal rotation with one impurity species has been treated with a moment approach, and an analytic expression for the neo-classical central impurity rotation toroidal rotation in purely Ohmic discharges is [273]

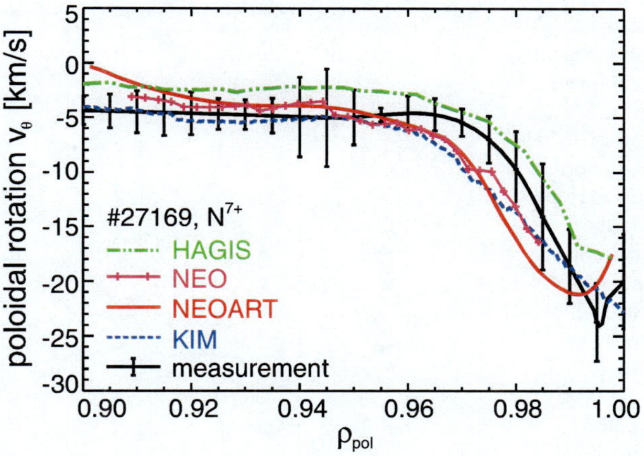

Fig. 4.2 Impurity ion poloidal rotation velocity measured on N^{7+} in a D plasma, along with the conventional neo-classical prediction in blue (dotted line) and the simulated profile using NEOART (red solid line), NEO (magenta crosses) and HAGIS (green dashed-dotted line). The measurements (black with error bars) were obtained in an ASDEX Upgrade H-mode plasma with $B_\phi = -2.5$ T, $I_p = 1$ MA, 2 MW ECRH, 10 MW NBI and a central line-averaged density of $8.1 \times 10^{19}/m^3$. Reproduced from Fig. 3 of [292] E. Viezzer *et al.* 'Evidence for the neoclassical nature of the radial electric field in the edge transport barrier of ASDEX Upgrade', Nucl. Fusion **54**, 012003 (2014), with the permission of the IAEA

Fig. 4.3 Comparison of measured v_θ profiles with neo-classical predictions for the thermal particles by NCLASS for two JET ITB discharges, with the measured v_θ values (stars) and the NCLASS v_θ predictions (solid lines). Reprinted with permission from Fig. 5 of [282] K. Crombé *et al.* 'Poloidal Rotation Dynamics, Radial Electric Field, and Neoclassical Theory in the Jet Internal-Transport-Barrier Region', Phys. Rev. Lett. **95**, 155003 (2005). Copyright (2005) by the American Physical Society

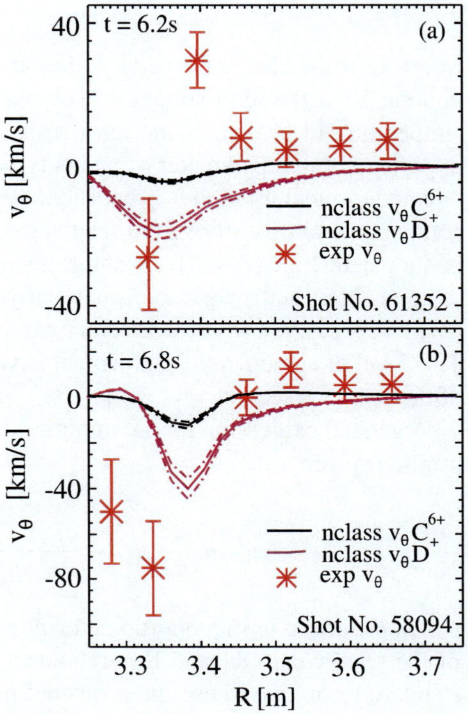

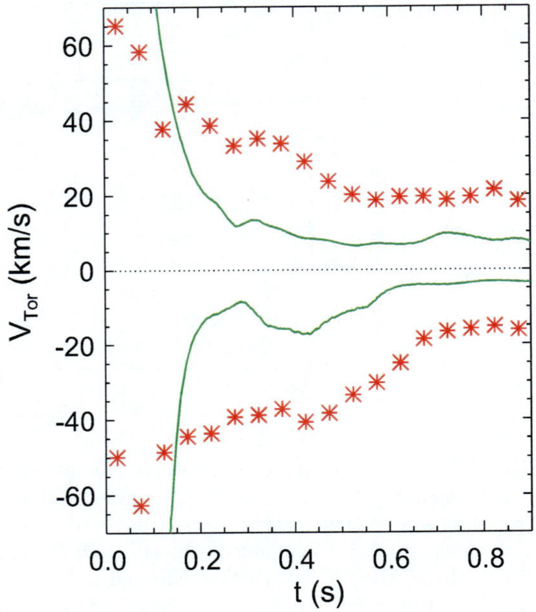

Fig. 4.4 Calculated neo-classical rotation velocity time histories (solid lines) compared with the measured values (asterisks) for C-Mod Ohmic plasmas with normal (bottom) and reversed (top) plasma currents. In the reversed plasma current configuration, the positive velocity indicates counter-current rotation. Reproduced from Fig. 7 of [34] J.E. Rice *et al.* 'X-ray observations of central toroidal rotation in ohmic Alcator C-Mod plasmas', Nucl. Fusion **37**, 421 (1997), with the permission of the IAEA

$$v_\phi^I = 4.19 \times 10^7 f \frac{Z_i}{\sqrt{\mu}} \frac{V_l}{R} \frac{T_i^{3/2}}{n_i} (cm/s) \tag{4.2}$$

where Z_i is the charge of the background ion, μ is the mass of the background ion in amu, V_l is the loop voltage in volts, R is the major radius in cm, T_i is the ion temperature in keV, n_i is the ion density in $10^{14}\,cm^{-3}$ and f is a function of the impurity strength parameter α, and is typically of order unity. This counter-current directed toroidal rotation is proportional to the loop voltage. This formula has been compared to observations from several devices, an example of which (from C-Mod) is shown in Fig. 4.4 [34]. In some instances, as in Fig. 4.4, there is qualitative (correct sign) and reasonable quantitative (within a factor of 2 or 3) agreement [21, 34, 298, 299] whereas in other examples the sign is wrong [12, 15, 19, 20]. This level of agreement/disagreement is very similar to the poloidal rotation results of the previous section.

A general expression for the impurity toroidal rotation across the complete radial profile is given as [273]

$$v_\phi^I = \frac{1}{B_\theta}\left[E_r - \frac{1}{en_i}\frac{\partial P_i}{\partial r} + \frac{(K_1 + \frac{3}{2}K_2)}{e}\frac{\partial T_i}{\partial r}\right] \tag{4.3}$$

which includes a term proportional to the radial electric field E_r, and where examples of the coefficients K_1 and K_2 are shown in Figs. 4.1 and 4.5, respectively. It is a standard technique to use the measured poloidal and toroidal velocities to infer E_r from force balance in this way. It is said that there is a chicken and egg relationship

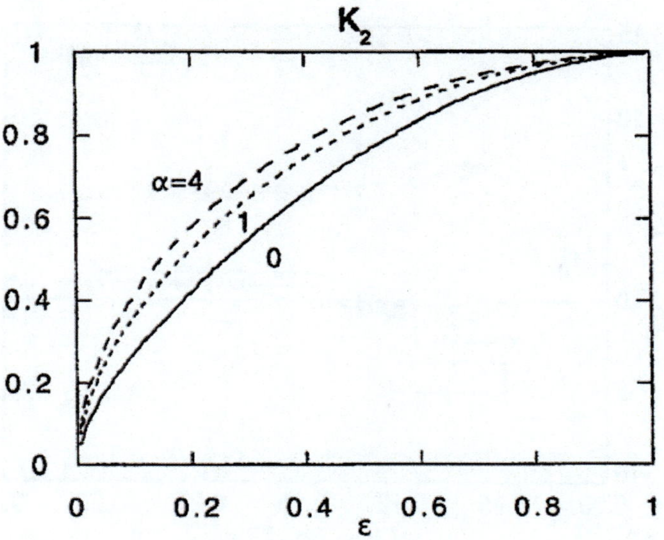

Fig. 4.5 The coefficient K_2 as a function of inverse aspect ratio ϵ in the ion banana regime for three values of the impurity strength parameter $\alpha = 0$ (solid), 1 (dotted) and 4 (dashed). Reproduced from Fig. 3 of [273] Y.B. Kim *et al.* 'Neoclassical poloidal and toroidal rotation in tokamaks', Phys. Fluids B **3**, 2050 (1991), with the permission of AIP Publishing

between **v**×**B** flows and E_r. Sometimes the poloidal rotation isn't measured and the neo-classical value is assumed; as was seen in the previous section, this is not always a safe approach. In some cases E_r is measured independently with a heavy ion beam probe and the toroidal rotation velocity can be compared directly to the neo-classical expression; qualitative agreement has been shown [17, 20]. Recently techniques have been developed to measure the background ion toroidal velocity directly [62, 84, 287, 290, 293, 300]. Since the difference between the ion and impurity toroidal velocity is independent of E_r, measured velocity differences can be directly compared with theory [62, 287, 290, 293]. The difference between the neo-classical toroidal impurity and ion angular frequencies [273] is

$$\omega_I - \omega_i = 3/4 K_2 (v_{th}\, \rho_{i\theta})/(R\, L_{T_i}) \tag{4.4}$$

where $\rho_{i\theta} = m_i \text{ c } v_{th}/(Z_i \text{ e } B_\theta)$, L_{T_i} is the ion temperature gradient scale length and K_2 is shown in Fig. 4.5. An example of clear disagreement is presented in Fig. 4.6 where the difference between deuterium and carbon toroidal velocities is shown as a function of collisionality for DIII-D Ohmic L-mode and ECH H-mode plasmas [290]. Clearly the sign and scaling are discordant. In some cases such as in the steep gradient regions of enhanced confinement modes, the normal ordering of neo-classical treatments [271–273] is violated. The sub-neo-classical approach [301], which includes finite Larmor radius effects, is more in line with the rotation measurements in some cases [302]. The takeaway message here is that while in some

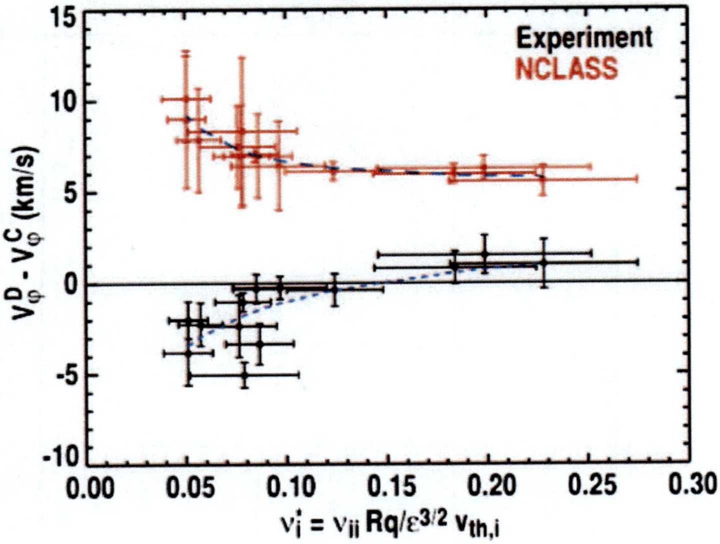

Fig. 4.6 Scaling of the differential toroidal rotation between deuterium and carbon of a range of collisionality for DIII-D plasmas. Reproduced from Fig. 3 of [290] B.A. Grierson *et al.* 'Collisionality scaling of main-ion toroidal and poloidal rotation in low torque DIII-D plasmas', Nucl. Fusion **53**, 063010 (2013), with the permission of the IAEA

instances agreement is good, in many cases the observed toroidal rotation deviates significantly from neo-classical predictions. There are indications that other effects, such as the influence of turbulence, must be taken into account, and this will be discussed in Chap. 5.

4.3 Poloidal Asymmetries in Edge Toroidal Rotation

Traditionally, edge rotation has only been measured on the low field side (LFS). A peculiarity in the edge toroidal velocity profile, the minimum between the pedestal top and LCFS, shown in Figs. 1.14 and 2.14, and routinely observed in DIII-D [184, 303] and ASDEX Upgrade [63] H-mode plasmas, is only found on the LFS. Recently, using high field side (HFS) gas puff charge exchange recombination spectroscopy [288, 291, 304–306], it is possible to investigate poloidal variation of edge flows. An example from an ASDEX Upgrade H-mode discharge [288] is shown in Fig. 4.7. As can be seen in the second frame, the toroidal impurity rotation just inside the top of the pedestal (bottom frame) is considerably higher on the HFS than on the LFS. This is due to the in-out density asymmetries and the effect on Pfirsch-Schlüter flows. Similar poloidal asymmetries have been predicted for KSTAR H-mode plasmas [307]. One *caveat* to the ASDEX Upgrade result is that the matching of the separatrix on the LFS and HFS was achieved by aligning

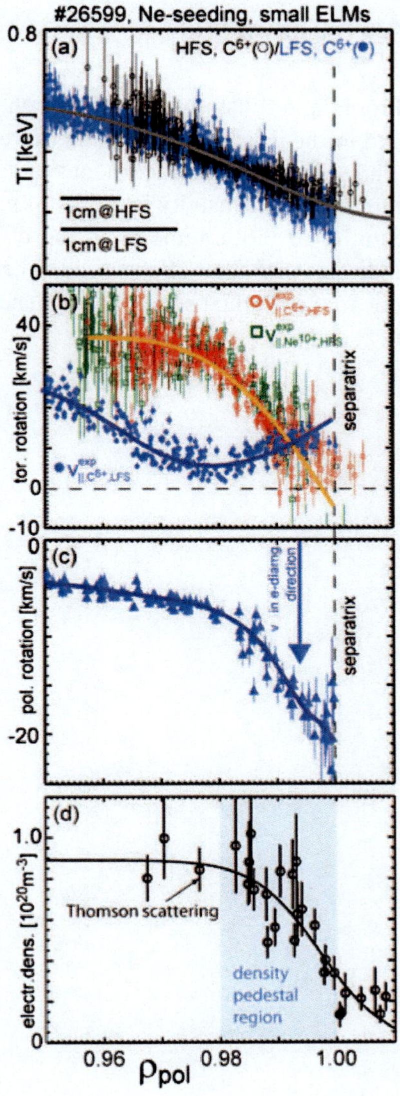

Fig. 4.7 For an ASDEX Upgrade H-mode plasma, (**a**) HFS/LFS meaurements of T_i to accomplish the alignment of the rotation profiles. (**b**) Toroidal/parallel rotation on the HFS/LFS. (**c**) Poloidal rotation of C^{6+} as measured via an independent, poloidal viewing geometry. (**d**) The electron density as measured by Thomson scattering is provided as an independent identification of the density pedestal region. The lines are spline fits to the data. Reproduced from Fig. 4 of [288] T. Pütterich *et al.* 'Poloidal asymmetry of parallel rotation measured in ASDEX Upgrade', Nucl. Fusion **52**, 083013 (2012), with the permission of the IAEA

the edge ion temperature profile, and similar results were found for C-Mod H-mode plasmas [305]. Alternatively if the edge E_r profiles were forced to align then there would be an in-out asymmetry of the edge ion temperature profile, which then must be explained. Interestingly, the edge E_r and T_i profiles were both simultaneously aligned in C-Mod I-mode discharges [305]. General information on I-modes may be found in [140, 308–316].

4.4 Comments

From the results presented in this chapter it is apparent that both measured poloidal and toroidal rotation sometimes agrees with the predictions of neo-classical theory while in other instances does not. As will be seen in Chap. 6, the observed momentum diffusivity is orders of magnitude larger than neo-classical predictions. Similar to the arguments in the chapters on momentum sources and sinks, this indicates that other effects need to be taken into account, such as the intrinsic torque due to the residual stress in turbulence dominated regimes. This is the subject of the next chapter.

Chapter 5
Intrinsic Rotation and the Residual Stress Π^{res}

Momentum sources and sinks are mostly offset by the Reynolds stress, which consists of three terms (see Eq. 2): diffusion, convection and the residual stress, Π^{res}. Diffusion and convection are strictly momentum transport coefficients which can affect the rotation profile shape and evolution, but cannot act as a momentum source. Multiple experiments have measured these [13, 14, 35, 36, 54, 71, 86–95, 97–104, 106, 109–118, 120, 191], and they will be covered in detail in Chap. 6. A unique feature of the third term is that the divergence of the residual stress constitutes an intrinsic torque and as such, functions as a momentum source. The residual stress is governed by the underlying turbulence [11], which can dominate over collisional neo-classical transport, can depend upon the density, temperature, pressure and current density profiles, and can change sign. This latter attribute plays a role in the rotation reversal phenomenon (see Sect. 5.2.3). Many examples will be provided in this section of how the rotation can be driven by changes in various gradients, as well as the collisionality and magnetic configuration, using external heating and current drive, and in Ohmic plasmas. In the absence of external momentum sources, if the initial velocity and velocity gradient are zero, the only way to spin up the plasma is through Π^{res} [11].

Evidence for the role of Π^{res} in driving intrinsic rotation has been steadily building. Examination of the velocity profile evolution following changes from co- to counter-current NBI in JFT-2M plasmas demonstrated that there is a finite momentum flux even at zero momentum gradient and a finite offset velocity gradient even at zero momentum flux [90, 91]. It was further shown that this off diagonal term in the momentum flux is proportional to the ion temperature gradient and inversely proportional to the plasma current [90, 91]. These properties are demonstrated in Fig. 5.1 [4]. Further evidence came by a completely different approach, from observations of intrinsic rotation in C-Mod plasmas [156, 157] which demonstrated the origin at the plasma periphery [35, 36] and scaling with the edge temperature gradient [140, 317] inversely proportional to the plasma current [157, 317] (Sect. 5.1.2).

J. Rice, *Driven Rotation, Self-Generated Flow, and Momentum Transport in Tokamak Plasmas*, Springer Series on Atomic, Optical, and Plasma Physics 119,
https://doi.org/10.1007/978-3-030-92266-5_5

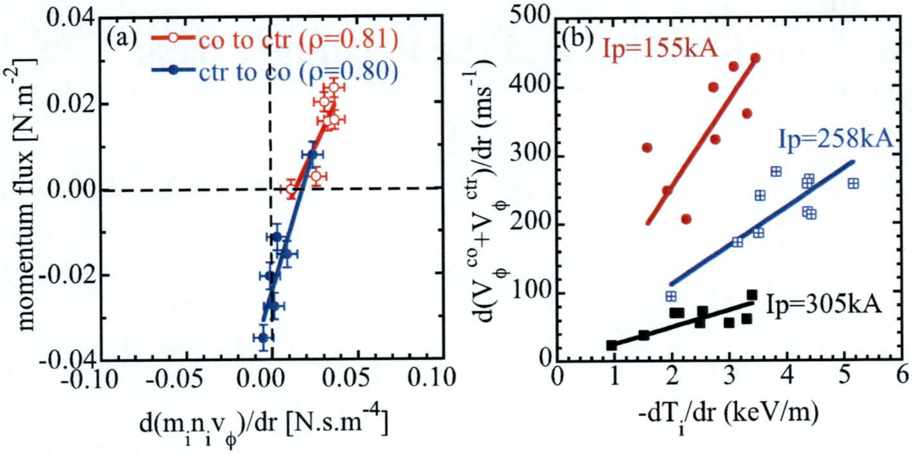

Fig. 5.1 (**a**) Radial flux of toroidal momentum as a function of the gradient of toroidal momentum for the discharges when the direction of the NBI is reversed from co- to counter- (open circles) and from counter- to co- (closed circles). (**b**) ITG dependence of offset rotation gradient. Reproduced from Fig. 24 of [4] K. Ida and J.E. Rice 'Rotation and momentum transport in tokamaks and helical systems', Nucl. Fusion **54**, 045001 (2014), with the permission of the IAEA

There are two broad classes of these experiments: intrinsic rotation in enhanced (Sect. 5.1) confinement regimes (H- and I-mode, and plasmas with internal transport barriers) and in L-mode (Sect. 5.2). I-mode is a confinement regime which exhibits the energy and momentum confinement of H-mode with the particle and impurity confinement of L-mode [140, 308–317]. In H- and I-mode plasmas, the intrinsic rotation is typically in the co-current direction, with Mach numbers as large as 0.3 in the core. This rotation is associated with the quality of energy confinement through global parameters such as the stored energy or H-factor, or local quantities such as pressure or temperature gradients. In contrast, intrinsic rotation in L-mode plasmas has a complicated dependence on magnetic configuration and collisionality, and can be in either the co- or counter-current direction, with dynamic rotation reversals sometimes observed.

5.1 Enhanced Confinement Regimes

Enhanced confinement regimes exhibit transport barriers: in the edge density and temperature profiles for high confinement mode (H-mode), in the edge temperature profile for improved confinement mode (I-mode) and in interior transport channels for ITB plasmas. This section will consider rotation in Ohmic H-modes (Sect. 5.1.1), ICRF H- and I-modes (Sect. 5.1.2), ECH H-modes (Sect. 5.1.3), NBI H-modes (Sect. 5.1.5) and in plasmas with ITBs (Sect. 5.1.6). Effects of the addition of ECH (Sect. 5.1.3) and LH (Sect. 5.1.4) to H-mode target plasmas are also summarized.

5.1.1 Ohmic H-Modes

Toroidal rotation in Ohmic H-modes will be considered first since without external momentum sources, the intrinsic torque across the L-H transition can be isolated. In Ohmic H-mode plasmas, there is no external momentum source other than the toroidal electric field, so that any change in the rotation at the L-H transition is entirely due to changes in confinement parameters. Co-current rotation changes following Ohmic L-H transitions have been observed in COMPASS-D [318], C-Mod [69, 70, 156, 157, 319, 320], DIII-D [184, 303] and NSTX [321]. An example is shown in Fig. 5.2 from C-Mod [70]. This H-mode was achieved by operating at relatively high plasma current (1.1 MA), in order to increase the total input power, and by ramping down the toroidal magnetic field from 5.4 to 4.3 T, in order to lower the H-mode threshold, which has been determined to scale as B_T. This plasma entered H-mode at 0.82 s, when the D_α signal dropped, and the plasma stored energy, the electron density (and temperature) and the central rotation velocity all increased. During the H-mode period, the plasma stored energy nearly doubled, reaching 105 kJ, the central electron density also doubled, increasing to $4.2\times10^{20}/m^3$, while the central toroidal rotation velocity exceeded 30 km/s. Here, a positive rotation velocity indicates co-current rotation.

The magnitude of the rotation change following the Ohmic L-H transition has been found to scale with the change in plasma stored energy [69, 70, 157, 160], and inversely proportional to the plasma current. This is demonstrated in Fig. 5.3, which shows the central impurity toroidal rotation velocity increase (the difference between the H-mode and pre-H-mode value) as a function of the stored energy increase, normalized to the plasma current, for a series of C-Mod Ohmic H-mode discharges. This is the same scaling observed in ICRF H-modes [156, 157, 160], and will be discussed in the next section.

5.1.2 ICRF H- and I-Modes

Similar toroidal rotation behavior has been observed in H-mode, I-mode and other enhanced confinement regimes formed with ICRF waves in JET, C-Mod, Tore Supra, KSTAR and EAST, where co-current changes in the toroidal rotation have been seen. The directionality of the intrinsic flow is demonstrated through a comparison of toroidal rotation time histories in C-Mod H-mode discharges with forward and reversed plasma current, as depicted in Fig. 5.4 [160]. Both plasmas had on-axis heating with comparable ICRF power levels (third panel) and subsequent stored energy increases (second frame); the changes in the rotation velocities (bottom panel) following the L-H transitions were of similar magnitude but in opposite directions, remaining co-current. The time histories of the rotation velocities were very similar to the plasma stored energy time histories, although for the reversed current case the rotation velocity was 'negative'. There was a delay,

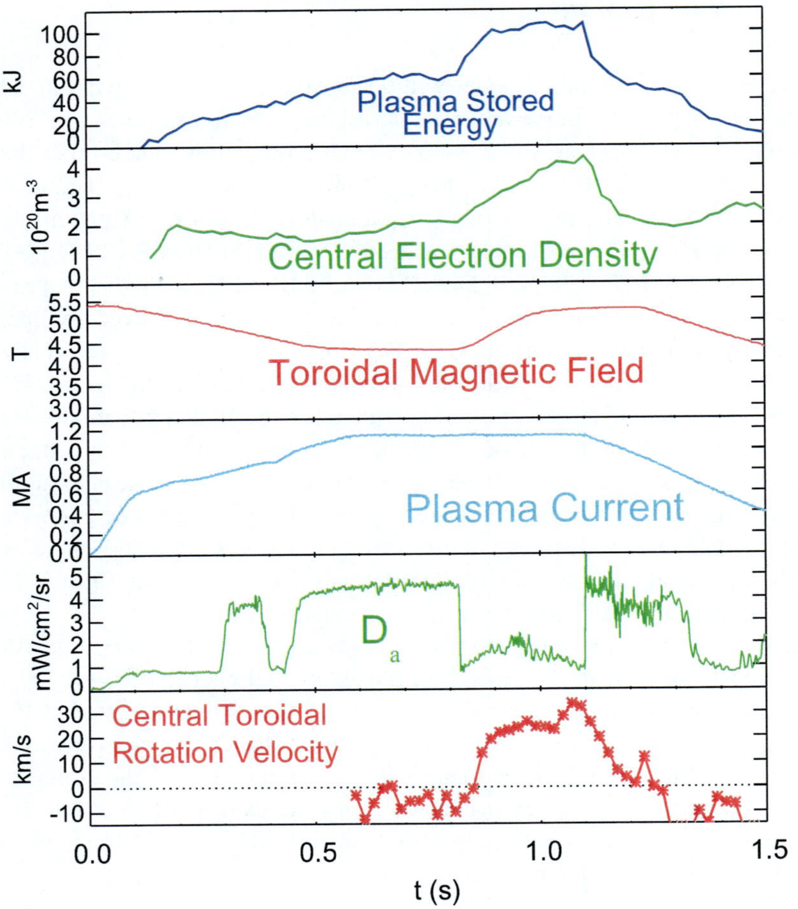

Fig. 5.2 Parameter time histories for a 1.1 MA Ohmic H-mode discharge. In the top frame is the plasma stored energy, in the second frame is the central electron density, in the third frame is the toroidal magnetic field, in the fourth frame is the plasma current and in the fifth frame is the D_α emission. In the bottom frame is the central toroidal rotation velocity of argon ions. Reproduced from Fig. 1 from [70] J.E. Rice et al. 'Impurity toroidal rotation and transport in Alcator C-Mod ohmic high confinement mode plasmas', Phys. Plasmas **7**, 1825 (2000), with the permission of AIP Publishing

about equal to the energy confinement time, of the modulations in the velocity traces relative to the stored energy signals. This delay is because the L-H transition begins in the edge pedestal region, and the appearance of the rotation in the plasma center takes a momentum confinement time. It should be noted that the toroidal magnetic field was reversed with the plasma current. Following the L-H or L-I transition, the velocity is observed to propagate from the pedestal region into the core [35, 36, 71, 140, 319]. This is demonstrated in Fig. 5.5 which shows the toroidal rotation velocity evolution at several radii following the L-I transition in a C-Mod plasma [140]. After the transition to I-mode (0.7 s), the co-current change in the

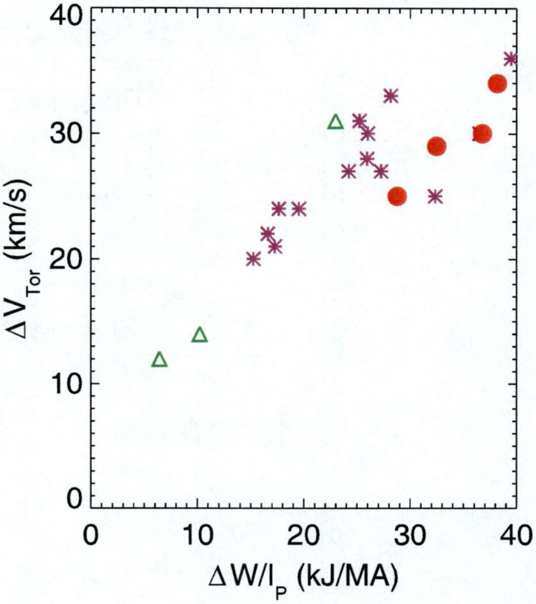

Fig. 5.3 The toroidal rotation velocity increase during the Ohmic H-mode as a function of the plasma stored energy increase, normalized to the plasma current, sorted by plasma current. Triangles—0.8 MA, asterisks—1.0 MA, dots—1.2 MA. Reproduced from Fig. 4 of [70] J.E. Rice et al. 'Impurity toroidal rotation and transport in Alcator C-Mod ohmic high confinement mode plasmas', Phys. Plasmas **7**, 1825 (2000), with the permission of AIP Publishing

toroidal rotation first appears at the plasma edge, then propagates in to the plasma center. This effect can be used to characterize momentum transport and will be discussed in Chap. 6. The magnitude of v_ϕ is correlated with global confinement parameters such as the plasma pressure [142, 159], stored energy [156, 157, 160, 161, 319, 320, 322, 323], confinement H factor [156, 158, 159] and β_N [320, 322], and is inversely proportional to the plasma current [157, 160, 317, 322]. The relation between co-current toroidal rotation (angular momentum density) and the ion pressure was first demonstrated in JET H-mode plasmas [142], and confirmed in Tore Supra enhanced confinement discharges [158]. As was seen in Figs. 5.2 and 5.4, the toroidal velocity and stored energy waveforms exhibit similar behavior. The correlation between these two is shown in Fig. 5.6 for C-Mod H-mode plasmas [160]. This is a comparison between the rotation velocity increase before and during ICRF and Ohmic H-modes and the plasma stored energy increase normalized to the plasma current, before and during ICRF and Ohmic H-modes. All of the ICRF data are for plasmas with the resonance close to the magnetic axis. The ICRF H-mode points are sorted by plasma current; there is a larger velocity increase for a given stored energy increase at lower plasma current. The Ohmic H-mode points display a similar trend (as was seen in Fig. 5.3), although the stored energy increases are modest since with only Ohmic power input, it is difficult to exceed the H-mode threshold. The Ohmic points have not been sorted by plasma current, but the range presented in the figure is from 0.8 to 1.4 MA. The inverse dependence of the toroidal rotation velocity on plasma current for C-Mod H-mode plasmas is demonstrated in Fig. 5.7 [157]. This shows the change in the central toroidal rotation velocity as a function of plasma current in H-mode discharges for a subset of the points in

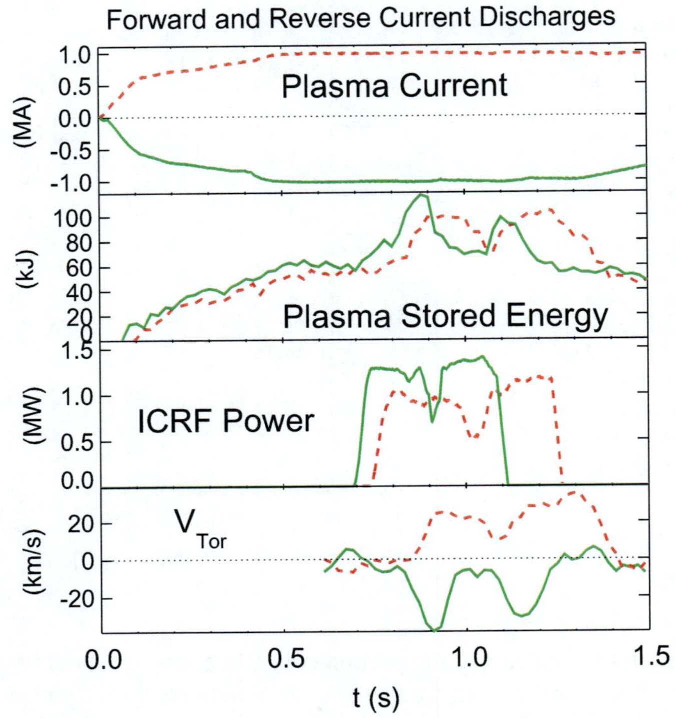

Fig. 5.4 A comparison of toroidal rotation during C-Mod ICRF heated H-mode plasmas with forward (red dashed lines) and reversed (solid green lines) plasma current. Reproduced from Fig. 1 of [160] J.E. Rice et al. 'Observations of impurity toroidal rotation suppression with ITB formation in ICRF and ohmic H mode Alcator C-Mod plasmas', Nucl. Fusion **41**, 277. (2001), with the permission of the IAEA

Fig. 5.6 with a change in the plasma stored energy between 40 and 60 kJ. Similar scalings have been seen in Ohmic (Sect. 5.1.1) and ECH (Sect. 5.1.3) [184, 322, 324] H-modes, so this behavior is universal, and indicates that it is not a direct ICRF wave effect. A comparison of results from C-Mod (ICRF and Ohmic H-modes), Tore Supra (ICRF enhanced confinement plasmas) [161] and DIII-D (Ohmic and ECH H-modes) [184] is shown in Fig. 5.8. There is a clear trend of the toroidal rotation increasing with stored energy (normalized to the plasma current) on all three devices, with the different slopes containing some machine size information [322]. These global scalings have been attributed to increases in the local pedestal pressure and temperature gradients. More recent JET results do not show these correlations [192].

Experiments in C-Mod have attempted to distinguish between the role of the pressure or temperature gradient by comparing the rotation behavior in H- and I-mode plasmas. I-mode [140, 308–316] is an enhanced confinement regime which exhibits H-mode-like energy confinement (through the edge temperature pedestal) but with L-mode-like particle and impurity confinement (without an edge density

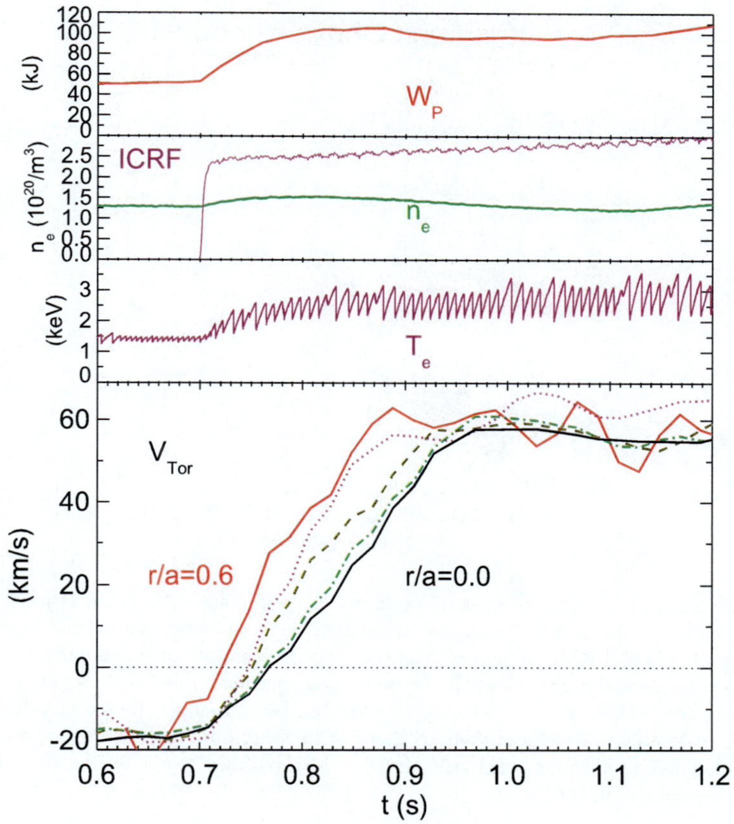

Fig. 5.5 Parameter time histories for an I-mode discharge. In the bottom frame is the normalized chord-averaged toroidal rotation velocity at several radial locations, evenly spaced from r/a=0.6 to r/a=0.0. Reprinted with permission from Fig. 3 of [140] J.E. Rice et al. 'Edge Temperature Gradient as Intrinsic Rotation Drive in Alcator C-Mod Tokamak Plasmas', Phys. Rev. Lett. **106**, 215001 (2011). Copyright (2011) by the American Physical Society

pedestal). With the lack of a density pedestal, the pedestal pressure gradient is lower in I-mode than in H-mode [313, 315], leading to an absence of ELMs, but with similar energy confinement characteristics (and temperature pedestals). For the same stored energy increase, the rotation in I-mode is very similar to that in H-mode [140]. Also similar to H-mode, the core toroidal rotation velocity in I-mode is observed to originate in the edge plasma, as seen in Fig. 5.5. These findings suggest that it is the edge temperature gradient which drives the rotation, and not the edge pressure gradient. This is demonstrated in Fig. 5.9 where the rotation changes in H- and I-mode are compared to the edge pressure and temperature gradients [140]. For similar ICRF power levels (bottom frame), the stored energy (top frame) and velocity (fourth frame) changes are quite similar. The changes in the pedestal temperature gradients are also similar, whereas there is a much larger change in the H-mode pedestal pressure gradient due to the contribution from the

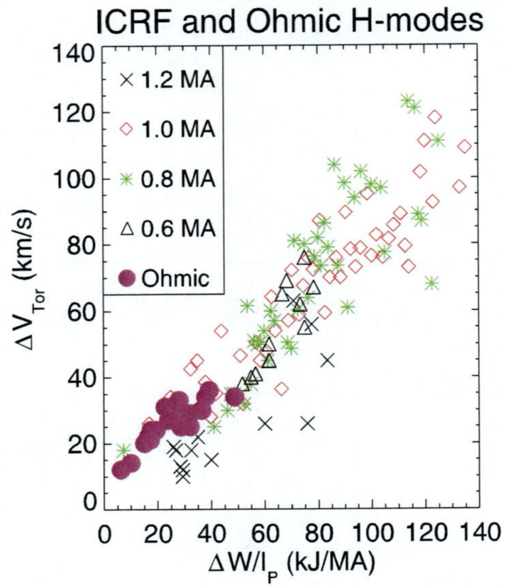

Fig. 5.6 The change in the toroidal rotation velocity (the difference between the H-mode and pre-H-mode values) as a function of the change in the plasma stored energy normalized to the plasma current, for Ohmic and ICRF H-modes. The Ohmic values are all shown as purple dots while the ICRF points have different symbols for the various plasma currents: black ×s for 1.2 MA, red diamonds for 1.0 MA, green asterisks for 0.8 MA and black triangles for 0.6 MA. Reproduced from Fig. 2 of [160] J.E. Rice et al. 'Observations of impurity toroidal rotation suppression with ITB formation in ICRF and ohmic H mode Alcator C-Mod plasmas', Nucl. Fusion **41**, 277. (2001), with the permission of the IAEA

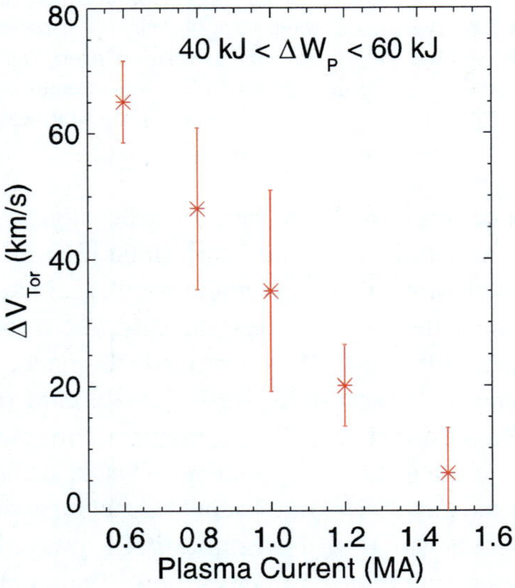

Fig. 5.7 The toroidal rotation velocity as a function of plasma current for stored energy increases between 40 and 60 kJ. Reproduced from Fig. 5 of [157] J.E. Rice et al. 'Central impurity toroidal rotation in ICRF heated Alcator C-Mod plasmas', Nucl. Fusion **39**, 1175. (1999), with the permission of the IAEA

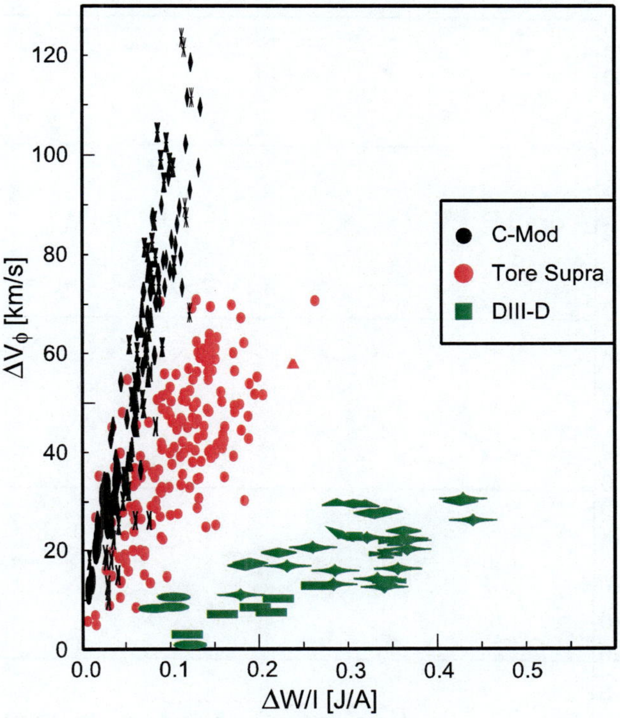

Fig. 5.8 The intrinsic rotation velocity (the difference between the L-mode velocity and the enhanced confinement value) as a function of the change in the stored energy normalized to the plasma current for Alcator C-Mod (black), Tore Supra (red) and DIII-D (green). Reproduced from Fig. 1 of [322] J.E. Rice et al. 'Inter-machine comparison of intrinsic toroidal rotation in tokamaks', Nucl. Fusion **47**, 1618. (2007), with the permission of the IAEA

pedestal density gradient. The magnitude of the velocity change is proportional to the pedestal temperature gradient change, which suggests that ∇T is the relevant parameter in driving the intrinsic rotation in H- and I-mode [140]. This point is further emphasized in Fig. 5.10, which shows the toroidal rotation velocity as a function of pedestal pressure (left) and temperature (right) gradients for C-Mod H- and I-mode plasmas. For the H-mode points, there is a very good correlation between ΔV and edge ∇P. However, the I-mode points do not coincide, because the rotation is the same while ∇P is lower than in H-mode (as seen in Fig. 5.9). The two sets of points overlay very well for the pedestal ∇T, as demonstrated on the right, suggesting that it is ∇T which is key in driving intrinsic rotation. The sign, magnitude and common scaling of the velocity in H- and I-mode with $\nabla T/B_\theta$ has been captured by examination of Π^{res} evaluated in ITG turbulence dominated regimes [140, 317, 325]. It has been argued that turbulence converts thermodynamic free energy to macroscopic flow [11, 326–328]. Considering the entropy production rate balance order by order [325] gives an expression for the velocity gradient $-\chi_\phi \langle V_\parallel \rangle' + \Pi^{res} = 0$ where χ_ϕ is the momentum diffusivity. Using a particular

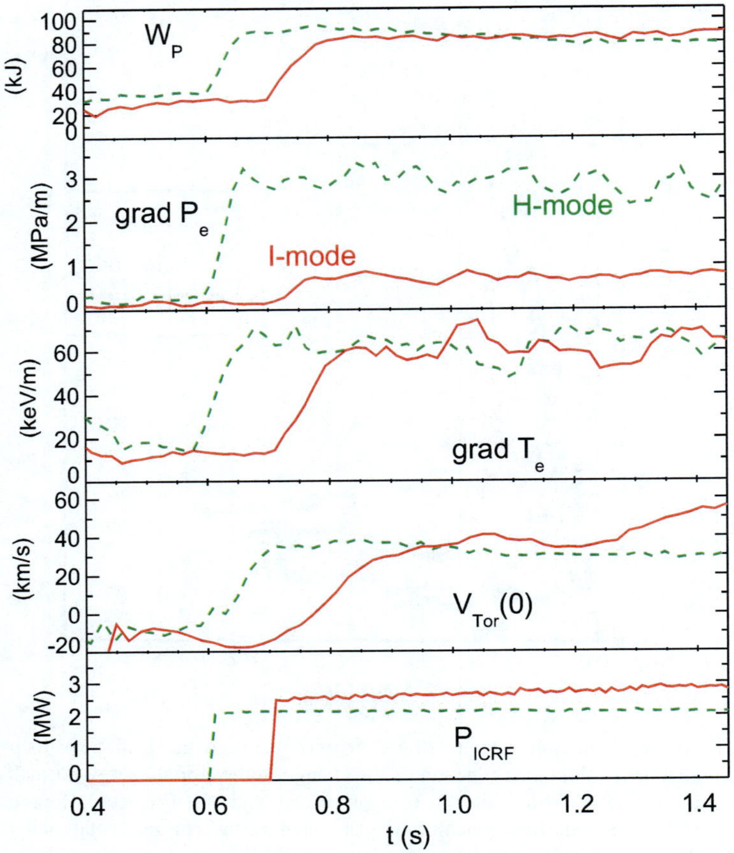

Fig. 5.9 Parameter time histories for C-Mod H-mode (green dashed) and I-mode (red solid) plasmas. From top to bottom, the plasma stored energy, pedestal pressure gradient, pedestal temperature gradient, central rotation velocity and ICRF power. Reprinted with permission from Fig. 1 of [140] J.E. Rice et al. 'Edge Temperature Gradient as Intrinsic Rotation Drive in Alcator C-Mod Tokamak Plasmas', Phys. Rev. Lett. **106**, 215001 (2011). Copyright (2011) by the American Physical Society

model of Π^{res} [325] derived considering ITG mode driven turbulence in a sheared cylinder with

$$\Pi^{res} = -\rho_* \frac{L_s}{2c_s} \chi_i \left(\frac{\nabla T}{T}\right)^2 v_{thi}^2 \tag{5.1}$$

where ρ_* is the normalized ion gyroradius (ρ_s/a), χ_i is the ion thermal conductivity, c$_s$ is the ion sound speed ($\sqrt{T_i/m_i}$) and L$_s$ is the magnetic shear scale length (R$_0$q/$\hat{s}$), leads (following integration) to an expression for the Mach number [325]

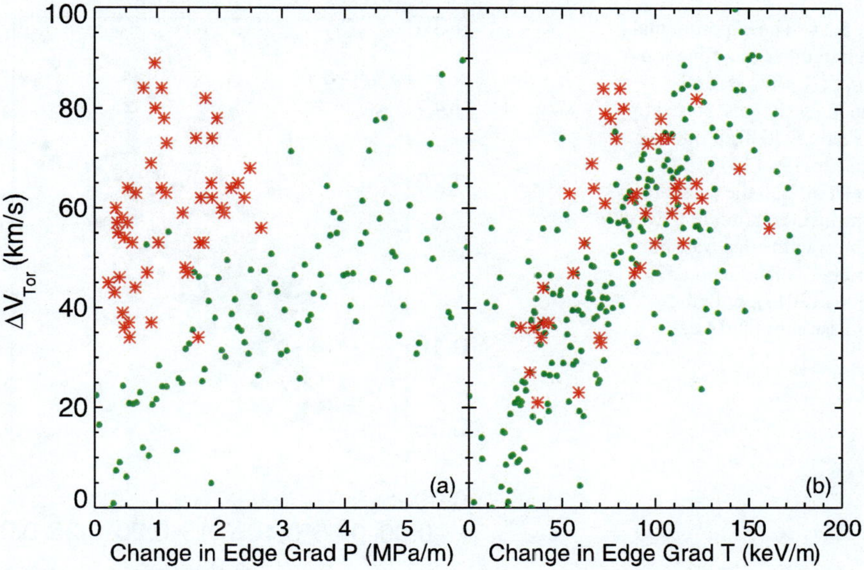

Fig. 5.10 The change in the core rotation velocity as a function of the change in the pedestal ∇P (**a**) and pedestal ∇T (**b**) for H-mode (green dots) and I-mode (red asterisks) plasmas. Reprinted with permission from Fig. 5 of [140] J.E. Rice et al. 'Edge Temperature Gradient as Intrinsic Rotation Drive in Alcator C-Mod Tokamak Plasmas', Phys. Rev. Lett. **106**, 215001 (2011). Copyright (2011) by the American Physical Society

$$M_i \cong \frac{1}{2}\rho_* \frac{\chi_i}{\chi_\phi} \frac{L_s}{L_T} \sqrt{\frac{T_i}{T_e}} \tag{5.2}$$

where L_T is the temperature gradient scale length (T/|∇T|). This model has been tested in C-Mod H- and I-mode plasmas with a range of toroidal magnetic fields between 2.7 and 7.8 T [317]. For C-Mod plasmas, edge ion and electron temperatures are similar, and the Prandtl number $P_r \equiv \chi_i/\chi_\phi$ is close to unity, so this reduces to $M_i \cong \frac{1}{2}\rho_* \frac{L_s}{L_T}$, which scales as ∇T q_{95}/B. From a database of over 300 discharges, the velocity was found to scale as $\nabla T^{0.8\pm0.3}$ $q_{95}^{1.0\pm0.2}$ / $B^{0.7\pm0.4}$, which captures these scalings. Comparison of the observed Mach number to the calculated $\frac{1}{2}\rho_* \frac{L_s}{L_T}$ is shown in Fig. 5.11. Values for different magnetic fields are interspersed. The agreement with the model is outstanding, over nearly an order of magnitude. This justifies the order by order entropy production rate balance paradigm, as well as the analytic model of the residual stress in the ITG turbulence dominated regime given by Eq. 5.1. The expression for the Mach number Eq. 5.2 works very well for C-Mod observations and it may be applied to future devices in order to get an estimate of the expected intrinsic rotation. For ITER Q = 5 and Q = 10 H-mode scenarios [329], with pedestal ρ_* values of ~0.001, L_T ~0.06 m and L_s ~8 m, the predicted Mach number is ~0.07, which corresponds to a core toroidal rotation velocity of ~50–75 km/s, depending on the core temperature and

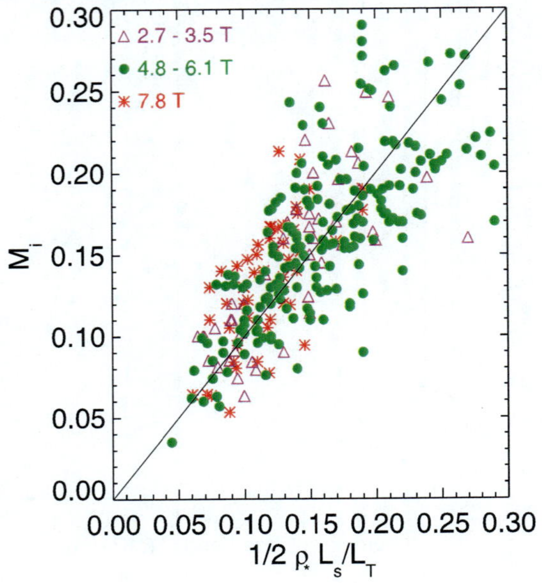

Fig. 5.11 The experimental Mach number as a function of $1/2\rho_* L_s/L_T$ for H- and I-mode discharges, sorted by magnetic field. Reproduced from Fig. 12 of [317] J.E. Rice et al. 'On the ρ_* scaling of intrinsic rotation in C-Mod plasmas with edge transport barriers', Nucl. Fusion **57**, 116004 (2017), with the permission of the IAEA

velocity peaking, and is large enough to suppress resistive wall modes (RWMs). This estimate is considerably lower (about a factor of 3) than those from a multi-machine scaling comparison [322], which did not account for the edge temperature gradient drive. Other reasons for this discrepancy are that ρ_* was evaluated in the core rather than in the pedestal, and all of the C-Mod points were at 5.4 T, with no accounting for L_s/L_T. For ARC [330] (a = 1.1 m, B = 9.2 T), with predicted pedestal $\rho_* = 0.0016$, $L_T = 0.025$ m and $L_s = 6$ m, the Mach number estimate is 0.2, which yields a substantial core velocity of over 200 km/s for a central temperature of 27 keV.

5.1.3 ECH H-Modes and Adding ECH

The toroidal rotation profile has been examined in L-mode target discharges with ECH [28, 184, 303, 324, 331–333], some of which entered H-mode. A common observation is that ECH H-modes exhibit co-current toroidal rotation in the edge regions [184, 303, 331]. This is illustrated in the top frame of Fig. 5.12, which shows a comparison of rotation frequency profiles for Ohmic L- and H-mode, and ECH H-mode discharges from DIII-D [184]. Comparing the target L-mode plasmas with the ECH H-mode cases (top frame) shows that the difference is co-current rotation outside of r/a = 0.5. The second frame demonstrates that 'core' ECH deposition only increases the rotation in the co-current sense in the outer regions, whereas 'off-axis' deposition leads to co-current rotation across the whole profile. Examination of the edge rotation (0.71 < r/a < 0.81) reveals a scaling with stored energy [184] similar

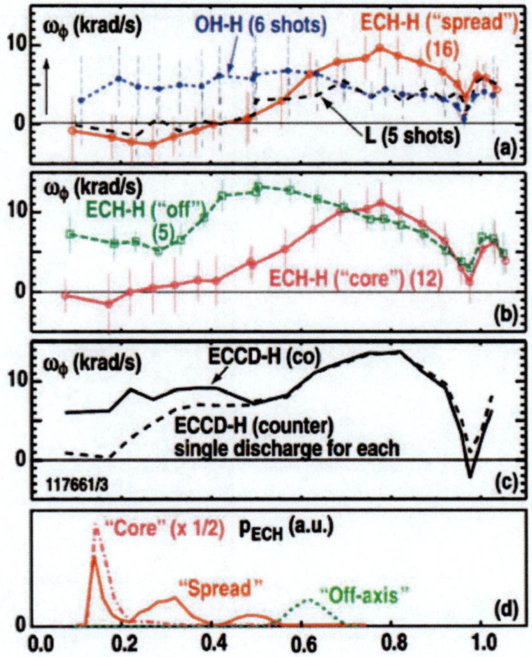

Fig. 5.12 Set-averaged rotation profiles, with m discharges in a set. (**a**) L-mode (pre-ECH) (dash, $m = 5$), Ohmic H-mode (dots, $m = 6$) 'spread' ECH H-mode (red open circles, $m = 16$, (**b**) 'core' ECH H-mode (red open circles, $m = 12$) and 'off-axis' ECH H-mode (green boxes, $m = 5$), (**c**) single examples with co- and counter-current ECCD H-mode with 'core' power deposition and (**d**) the ECH power deposition profiles for 'spread' (red solid line), 'core' (red dash-dot) and 'off-axis' (green dotted). Reproduced from Fig. 2 of [184] J.S. deGrassie et al. 'Intrinsic rotation in DIII-D', Phys. Plasmas **14**, 056115 (2007), with the permission of AIP Publishing

to that seen in Ohmic (Sect. 5.1.1) and ICRF (Sect. 5.1.2) enhanced confinement plasmas [322], regardless of ECH deposition location. This is shown in Fig. 5.13 for these DIII-D discharges [184]. Similar to the results in Ohmic (Fig. 5.3) and ICRF (Fig. 5.6) H-modes, the intrinsic rotation in ECH H-modes increases with stored energy normalized to the plasma current. These same DIII-D points were shown in Fig. 5.8 for comparison with C-Mod and Tore Supra observations. A different characteristic feature of some ECH H-mode plasmas compared to Ohmic and ICRF H-modes is the counter-current toroidal rotation observed in the core regions [184, 303, 333], an effect which depends on the ECH power deposition location. Off-axis power deposition leads to co-current rotation in the core [184], as seen in the second frame of Fig. 5.12. A dimensionless scaling study has been performed for DIII-D ECH H-mode plasmas [334] and the intrinsic Alfvén Mach number was found to scale as $M_A \sim \beta_N \rho_*$. This scaling also matches well with C-Mod and JET data [334].

Many of these same features are observed when ECH power is added to NBI H-mode plasmas. It is widely observed that when on-axis ECH is applied to

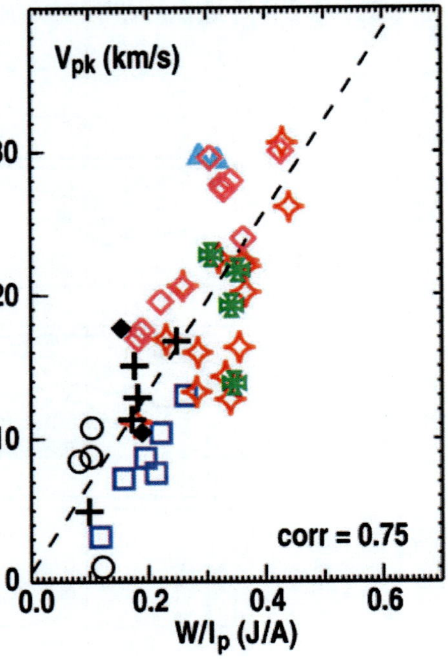

Fig. 5.13 Scaling of the rotation between 0.71 < r/a < 0.81 with stored energy normalized to the plasma current. Colored points are from ECH H-modes. Reproduced from Fig. 8 of [184] J.S. deGrassie et al. 'Intrinsic rotation in DIII-D', Phys. Plasmas **14**, 056115 (2007), with the permission of AIP Publishing

H-mode plasmas there is a counter-current change in the core toroidal rotation [115, 332, 335–340]. An example of this effect is shown in Fig. 5.14 for three ASDEX-Upgrade discharges [335]. A comparison of the red profiles indicates that on-axis ECH power deposition leads to a counter-current increment in the core rotation profiles. As the ECH power deposition is moved radially outward, the counter-current changes are reduced. This is a general result, regardless of the initial velocity of the target plasma. In JT-60U NBI H-mode plasmas where the ambient rotation velocity can be close to zero, application of on-axis ECH power also leads to counter-current changes in the core rotation, as can be seen in Fig. 5.15 [337]. Following the initiation of ECH at 7 s there was a core toroidal rotation increment in the counter-current direction. There was also an increase in the core electron temperature gradient and a decrease in the T_i gradient. A peaking of the electron density profile is commonly observed in conjunction with the flattening/hollowing of the velocity profile [115, 335]. Another common feature of observations in H-mode plasmas is that off-axis ECH power leads to co-current rotation in the core [335].

In some cases there is evidence that application of ECH causes the ambient rotation to approach an offset value [332, 336], similar to what was presented in Sect. 3.2. For co-rotating NBI plasmas, addition of ECH leads to a counter-current velocity increment while in counter-rotating NBI target plasmas the velocity change is in the co-current direction [332]. Another important variable in determining the rotation response to ECH is the effect of the q profile [333, 339].

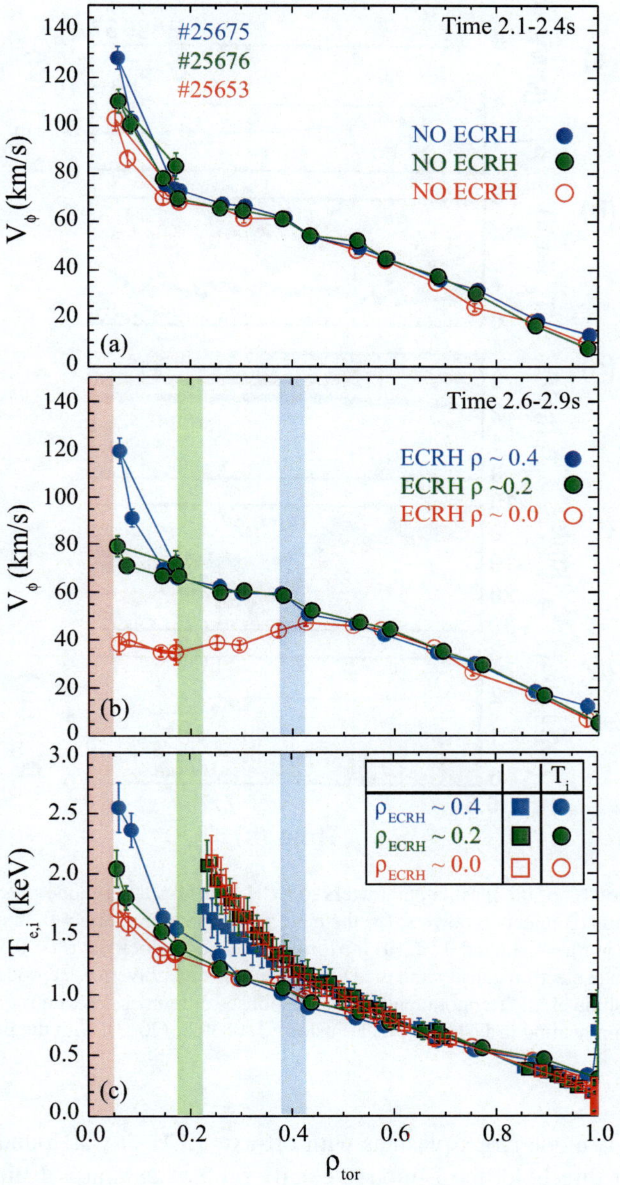

Fig. 5.14 Rotation and temperature profiles from three discharges in which 1.1 MW of ECRH power was scanned shot to shot from $0.4 > \rho_{Tor} > 0.0$. (**a**) The initial rotation profiles (no ECRH) from the three discharges overlay well indicating similar starting conditions. (**b**) The final rotation profiles (with ECRH) show a clear dependence on the ECRH power deposition location. (**c**) The ion temperature profiles (circles) show a small but distinct change with the ECRH deposition radius, while the electron temperature profiles (squares) are inconclusive due to incomplete profiles. Fig. 3 from [335] R.M. McDermott et al. 'Effect of electron cyclotron resonance heating (ECRH) on toroidal rotation in ASDEX Upgrade H-mode discharges', Plasma Phys. Control. Fusion **53**, 035007 (2011). ©IOP Publishing. Reproduced with permission. All rights reserved

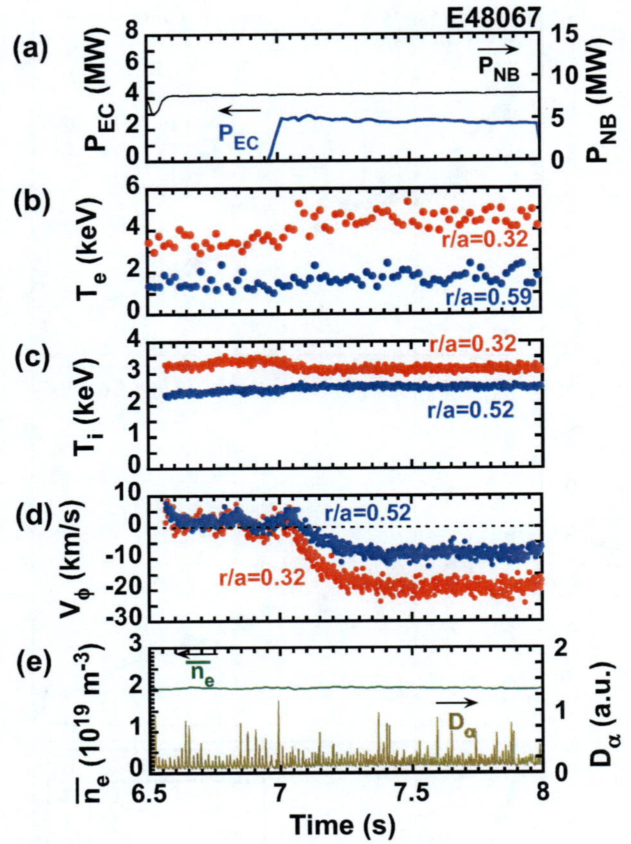

Fig. 5.15 Responses of the plasma parameters to ECH in a JT-60U H-mode plasma. Waveforms of (**a**) the EC and NB injection powers; (**b**) the electron temperature at r/a = 0.32 and 0.59; (**c**) the ion temperature at r/a = 0.32 and 0.52; (**d**) the toroidal rotation velocity at r/a = 0.32 and 0.52; and (**e**) the line average electron density and the D_α signal at the outer divertor. Reproduced from Fig. 1 of [337] M. Yoshida et al. 'Temporal and spatial responses of temperature, density and rotation to electron cyclotron heating in JT-60U', Nucl. Fusion **53**, 083022 (2013), with the permission of the IAEA

In DIII-D L-mode target plasmas with off-axis ECH, it was found that above a certain power threshold, the L-mode velocity profiles exhibit a distinct hollowing [341]. This is demonstrated in Fig. 5.16 which shows the main ion rotation frequency profiles for three different off-axis ECH power levels. The hollow rotation profile is quantitatively reproduced by global non-linear gyro-kinetic simulations. Above a critical heating power, the plasma becomes linearly unstable to ITG modes. Global non-linear simulations with the GTS code [342, 343] confirm a residual stress in the unstable region. ITG mode turbulence produces a negative instrinsic torque at the inner radii, leading to counter-current rotation and a positive torque outside generating co-current flow.

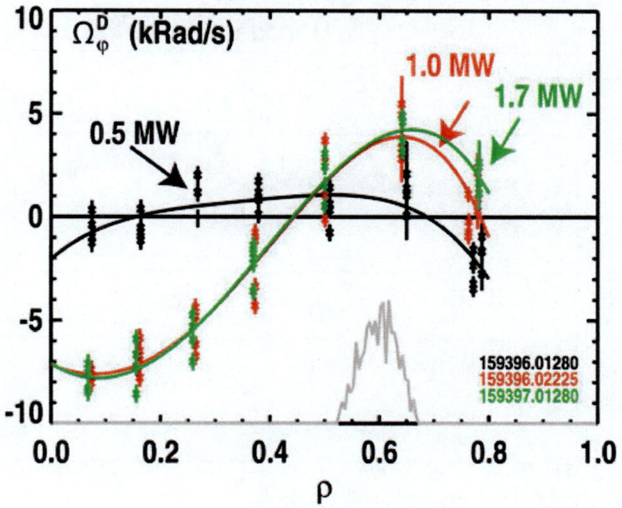

Fig. 5.16 Experimental main ion toroidal rotation profiles displaying a reversed rotation profile upon increasing the heating power from 0.5, 1.0 and 1.7 MW of ECH. Reprinted with permission from Fig. 1 of [341] B.A. Grierson et al. 'Main-Ion Intrinsic Toroidal Rotation Profile Driven by Residual Stress Torque from Ion Temperature Gradient Turbulence in the DIII-D Tokamak', Phys. Rev. Lett. **118**, 015002 (2017). Copyright (2017) by the American Physical Society

5.1.4 H-Modes with LH

Adding LH power to H-mode target plasmas can lead to a counter-current rotation change in conjunction with a peaking of the electron density profile [168, 169, 344], similar to what was observed with ECH, as was presented in the previous section. Shown in Fig. 5.17 are the time histories of parameters of interest for a C-Mod H-mode discharge with the addition of modest LH power [169]. There is a dramatic counter-current change in the central toroidal rotation (with little change in the outer regions), which is correlated with an increase in the electron density profile peaking, following application of a relatively small level of LH power. It is also remarkable that at this electron density, LH wave penetration is expected to be poor, and the exact mechanism for these effects is unknown.

5.1.5 NBI H-Modes

In spite of the large input torque from NBI (Sect. 2.1.1) information on intrinsic rotation in NBI H-mode plasmas can still be gleaned from balanced beam operation or through modulation experiments. The first indication of this was from experiments on JFT-2M which demonstrated a finite momentum flux at zero momentum gradient and a finite offset velocity gradient even at zero momentum flux [90, 91], as was

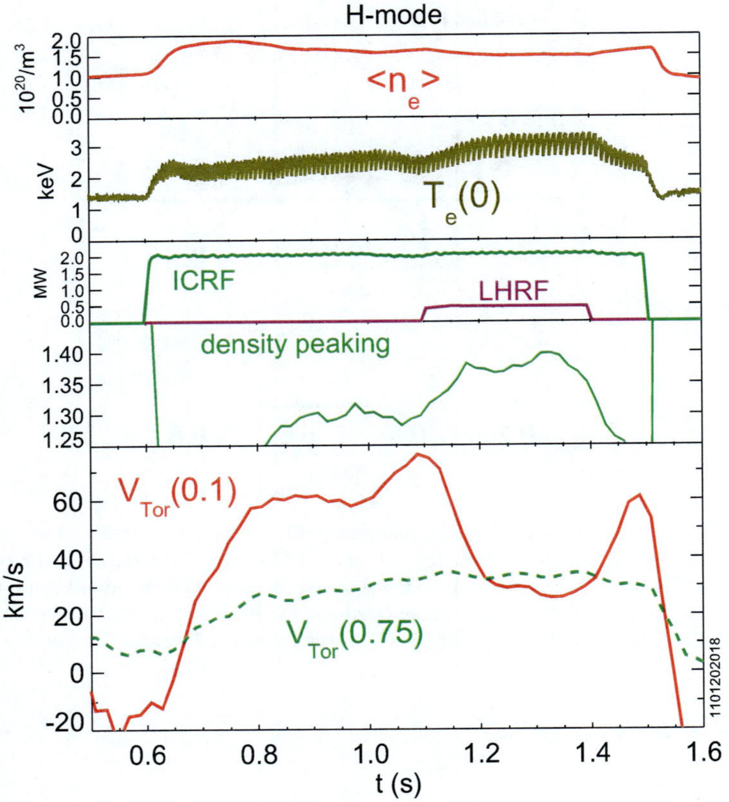

Fig. 5.17 Time histories of the line averaged electron density, central electron temperature, ICRF and LHRF power, electron density peaking (ratio of core to edge) and rotation velocity in the core (solid) and edge (dashed) for a 5.4 T, 0.62 MA (q_{95} = 6.3) USN C-Mod H-mode discharge. Reproduced from Fig. 20 of [169] J.E. Rice et al. 'Effects of LHRF on toroidal rotation in Alcator C-Mod plasmas', Nucl. Fusion **53**, 093015 (2013), with the permission of the IAEA

discussed in the begining of this chapter. This off diagonal term in the momentum flux was found to be proportional to the ion temperature gradient and inversely proportional to the plasma current [90, 91], similar to the deductions from ICRF heated H- and I-modes summarized in Sect. 5.1.2. Related experiments from DIII-D, where the input NBI torque from both co- and counter-current oriented neutral beam sources was varied shot to shot, reveal that there is a finite co-current velocity even with zero net torque. This is shown in Fig. 5.18 which shows the core rotation velocity as a function of total input torque [100]. At zero input torque, the core velocity exceeds 100 km/s; in order to reduce the velocity to zero, about 2 Nm of torque in the counter-current direction is required. Another way of demonstrating this is to show how much counter-current torque is required to offset the intrinsic co-current rotation. This is presented in Fig. 5.19 which shows a flat rotation profile near zero across the entire plasma radius, which was achieved with net counter-

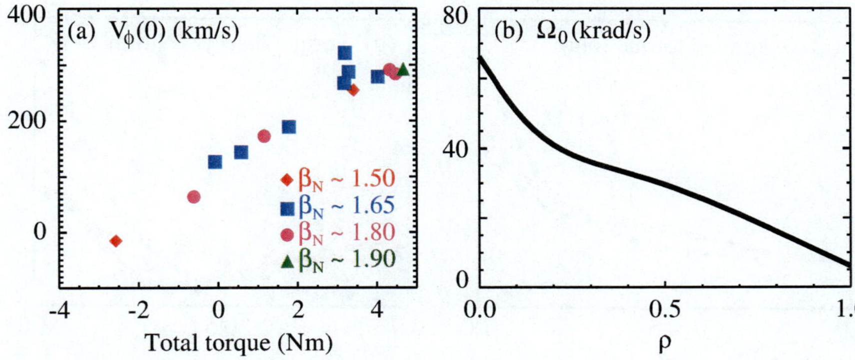

Fig. 5.18 (**a**) Central toroidal velocity versus total integrated torque. (**b**) Estimated intrinsic rotation profile, obtained from torque scan by interpolating the measured profiles to zero total integrated torque. Fig. 1 from [100] W.M. Solomon et al. 'Momentum confinement at low torque', Plasma Phys. Control. Fusion **49**, B313 (2007). ©IOP Publishing. Reproduced with permission. All rights reserved

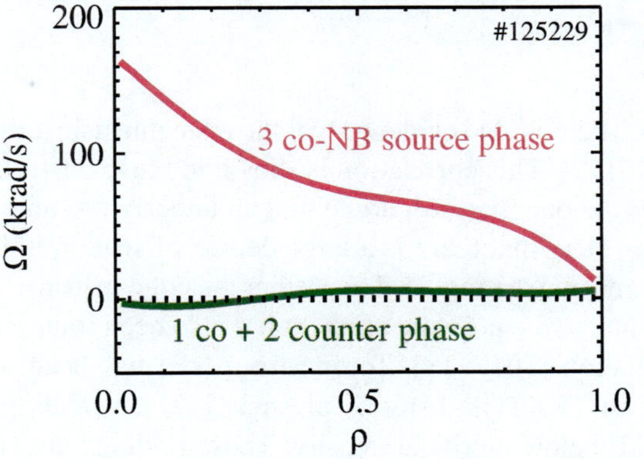

Fig. 5.19 Rotation frequency profiles with one co- and two counter- current neutral beam sources. With one net counter source applied, the rotation is effectively zero across the profile. For reference, the three co-NBI source phase is shown. Figure 3 from [100] W.M. Solomon et al. 'Momentum confinement at low torque', Plasma Phys. Control. Fusion **49**, B313 (2007). ©IOP Publishing. Reproduced with permission. All rights reserved

current input torque. Around 2.5 MW of counter-current NBI power was required to zero out the velocity profile. From the torque scan data, the intrinsic torque profile can be obtained by the following method. At each radius, the integrated angular momentum to that radius can be plotted against the integrated neutral beam torque to that radius. The x-intercept at each radius corresponds to the amount of beam torque required to cancel the intrinsic flow and bring the angular momentum to zero. The intrinsic torque determined in this fashion is shown in Fig. 5.20 [100]. The intrinsic torque is peaked near the plasma edge [110, 113, 197]; this is consistent with the results of Fig. 5.5 which indicate that intrinsic rotation originates at the

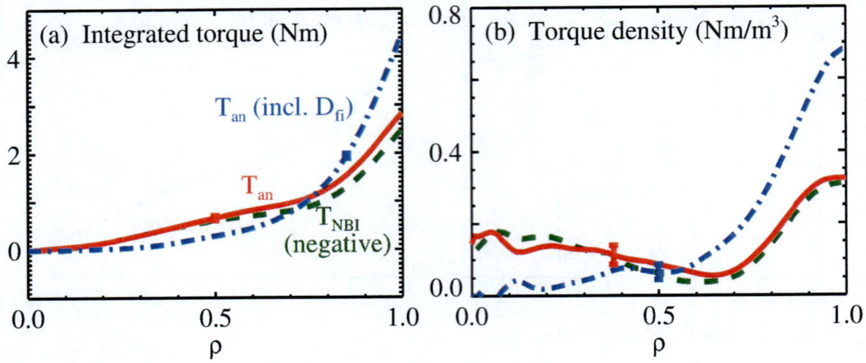

Fig. 5.20 (**a**) Intrinsic torque profile as approximated by the negative of the neutral beam torque when v_ϕ is close to zero across the whole profile (dashed) and is interpolated from the torque scan (solid). Including the effects of anomalous fast ion transport to match the measured neutron rate and stored energy modifies the inferred anomalous torque profile (dash-dot). (**b**) Inverted intrinsic torque density profiles. Figure 4 from [100] W.M. Solomon et al. 'Momentum confinement at low torque', Plasma Phys. Control. Fusion **49**, B313 (2007). ©IOP Publishing. Reproduced with permission. All rights reserved

plasma boundary. There is also evidence that the edge intrinsic torque scales with the pedestal ∇P [113, 197]. This correlation is shown in Fig. 5.21 from DIII-D H-mode plasmas [197]. While one may still argue that an underlying trend still exists in this larger dataset, it is clear that there is a large degree of scatter, indicating additional physics mechanisms may be required to predict the edge instrinsic torque. Evidence for Π^{res} proportional to the pressure gradient has also been found in JT-60U plasmas using NBI modulation [101, 119]. Torque from NBI has been used to cancel the intrinsic rotation in TEXTOR L-mode plasmas [121], revealing a residual stress proportional to ∇P below a critical density. These findings are consistent with the results from ICRF plasmas (Sect. 5.1.2) obtained in a completely different fashion.

The intrinsic torque has been determined over a large operational range in DIII-D, which allows for parameter studies in dimensionless quantities such as the normalized gyro-radius ρ_* and the collisionality ν_*. Shown in Fig. 5.22 is the edge intrinsic torque normalized to the ion temperature as a function of ρ_* for DIII-D H-mode plasmas [345]. There is a strong decrease in the intrinsic torque, with a $\rho_*^{-1.5}$ dependence. Using this plot of the intrinsic torque (normalized to the ion temperature), one may estimate the intrinsic torque for the expected ρ_* in ITER. Making assumptions about the momentum transport coefficients yields a rotation frequency of 20 kR/s [346, 347], similar to estimates presented in Sect. 5.1.2.

In an effort to reveal the underlying mechanism giving rise to the intrinsic torque, fluid turbulent stresses have been directly measured in the edge of DIII-D H-mode plasmas using a reciprocating Langmuir probe [76, 77]. A ubiquitous narrow co-current toroidal rotation layer at the separatrix is found in H-mode plasmas formed with co- and counter-current NBI (as well as with pure ECH). Correlations between toroidal and radial velocity fluctuations inside of this layer imply a significant

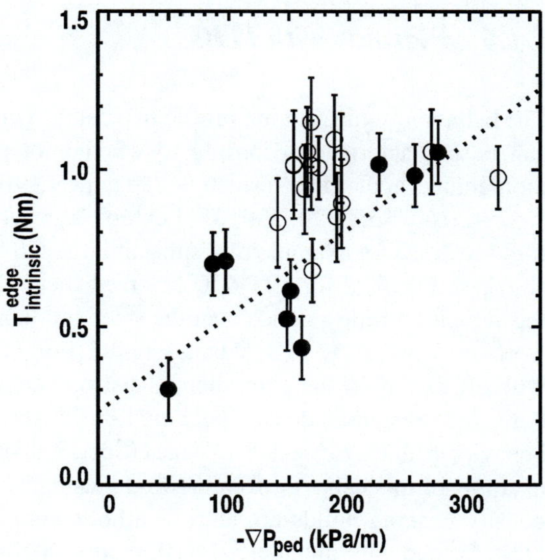

Fig. 5.21 Expanded dataset of edge intrinsic torque measurements versus pedestal pressure gradient. The original data corresponding to dedicated power and current scans are shown with solid symbols. Reproduced from Fig. 4 of [197] W.M. Solomon et al. 'Characterization of intrinsic rotation drive on DIII-D', Nucl. Fusion **51**, 073010 (2011), with the permission of the IAEA

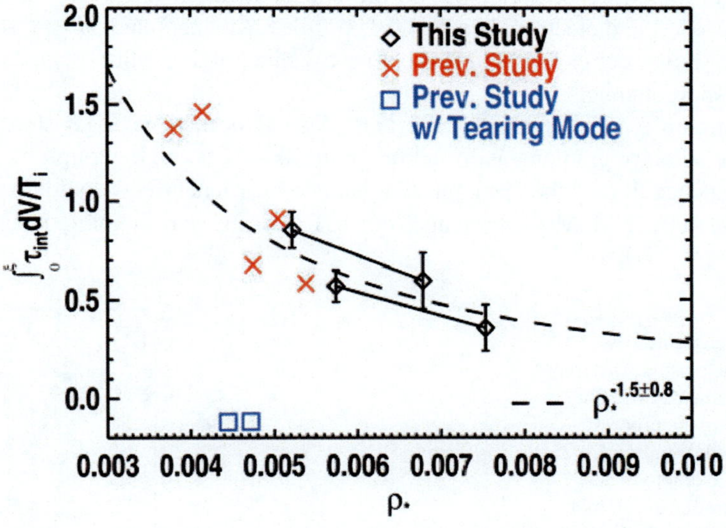

Fig. 5.22 The normalized intrinsic torque as a function of ρ_*. Reproduced from Fig. 9 of [345] C. Chrystal et al. 'Dependence of intrinsic torque and momentum confinement on normalized gyroradius and collisionality in the DIII-D tokamak', Phys. Plasmas **24**, 042501 (2017), with the permission of AIP Publishing

counter-current torque, which suggests that kinetic stresses may be important in driving the co-current rotation in the core. Recently, there is evidence that the preferential orbit loss of counter-current directed ions produces a co-current intrinsic rotation that peaks just inside the separatrix [199, 203].

5.1.6 Plasmas with ITBs

The behavior of the rotation profile in plasmas with internal transport barriers (ITBs) can be characterized according to whether or not there was substantial external momentum input. ITBs formed with balanced NBI [348, 349] or with off-axis ICRF [35, 36, 160, 163, 320, 350, 351] exhibit a negative velocity gradient near the ITB foot location. An example of a substantial notch in the velocity profile [348, 349] is shown in Fig. 5.23 for a JT-60U reversed shear plasma. The large negative notch in the velocity profile extends over the steep gradient ITB region. Similar observations were seen in TFTR ITB plasmas [348]. For ITBs produced in H-mode plasmas with off-axis ICRF heating, there is a strong counter-current velocity change in the core, and substantial density peaking [35, 36, 160, 163, 320, 350, 351], similar to the behavior with ECH (Sect. 5.1.3), LH (Sect. 5.1.4) and even in Ohmic ITBs [160]. An example of this behavior for a C-Mod plasma is shown in Fig. 5.24. As the electron density peaking builds up, there is a hollowing of the central velocity profile. For ITBs formed with unbalanced NBI, there is a positive break in slope of the toroidal velocity profile which corresponds to the barrier location in the ion temperature [293, 352–355] and electron density [354] profiles. An example from a DIII-D ITB plasma is shown in Fig. 5.25 [354]. There are distinct high gradient regions in these four transport channels.

A common feature of all of these ITB plasmas is that the **E**×**B** shearing rate (from the velocity gradient) is in the range of 10^5–10^6/s, high enough to suppress ITG turbulence. It may then be expected that neo-classical effects could dominate in the region of the ITB. Measurements from JET indicate there is not good agreement

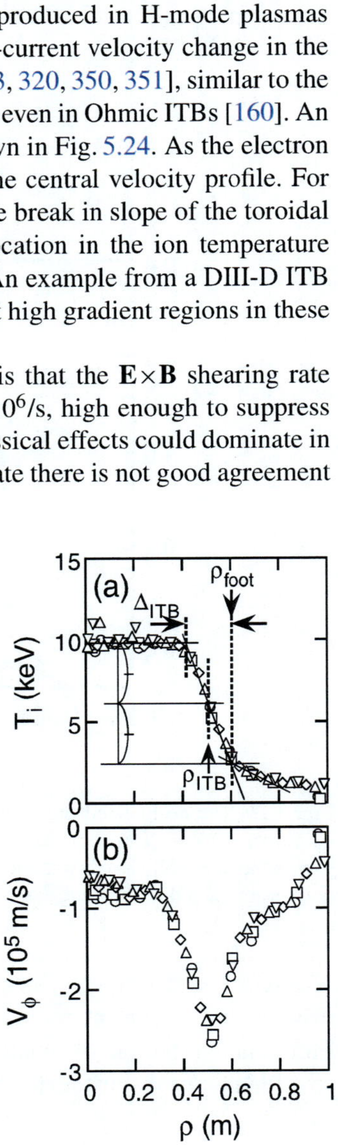

Fig. 5.23 Profiles of (**a**) T_i and (**b**) v_ϕ obtained by jogging the plasma. Different symbols indicate different times during the process. Reproduced from Fig. 1 of [349] Y. Sakamoto et al. 'Characteristics of internal transport barriers in JT-60U reversed shear plasmas', Nucl. Fusion **41**, 865 (2001), with the permission of the IAEA

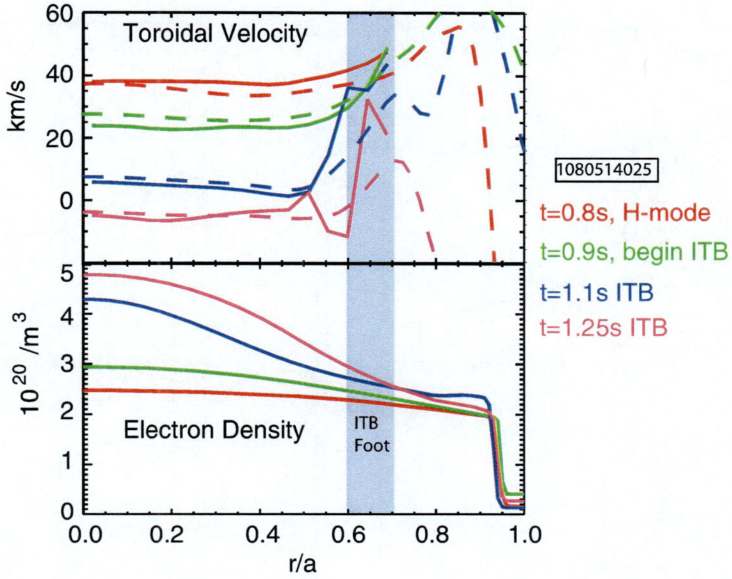

Fig. 5.24 Toroidal velocity profiles at four times in the development of an ITB compared with the electron density profiles: following the H-mode transition (red), beginning of peaking for the ITB (green), established ITB (blue), just prior to the back transition (violet). Solid lines are measured profiles. Dashed lines are smoothed profiles output from TRANSP/NCLASS using experimental data from $0 < r/a < 0.7$. Reproduced from Fig. 7 of [351] C.L. Fiore et al. 'Rotation and transport in Alcator C-Mod ITB plasmas', Nucl. Fusion **50**, 064008 (2010), with the permission of the IAEA

between measured impurity poloidal rotation profiles [195, 282] and neo-classical predictions, as was seen in Fig. 4.3. Recently techniques have been developed to measure directly the main ion rotation profiles [287, 290, 293]. Fair agreement between observed ion v_θ profiles and neo-classical calculations for the region just inside the ITB foot location can be seen in Fig. 5.26 for a DIII-D ITB plasma [293]. In other instances there is poor agreement, as was seen in Chap. 4. Another area that should be explored more in detail is the influence of the magnetic shear on the rotation profile in ITB plasmas [356].

5.1.7 Comments

There is extensive evidence that the co-current intrinsic rotation observed in enhanced confinement regimes on many devices can be tied to the temperature gradient with an inverse dependence on B_θ or plasma current [91, 140, 317]. There are several theoretical treatments of this problem that reproduce these results. Considering a neo-classical quasi-linear theory of fluctuation induced toroidal stress [357] captures the $1/B_\theta$ scaling with a Mach number of order 0.1. The accretion

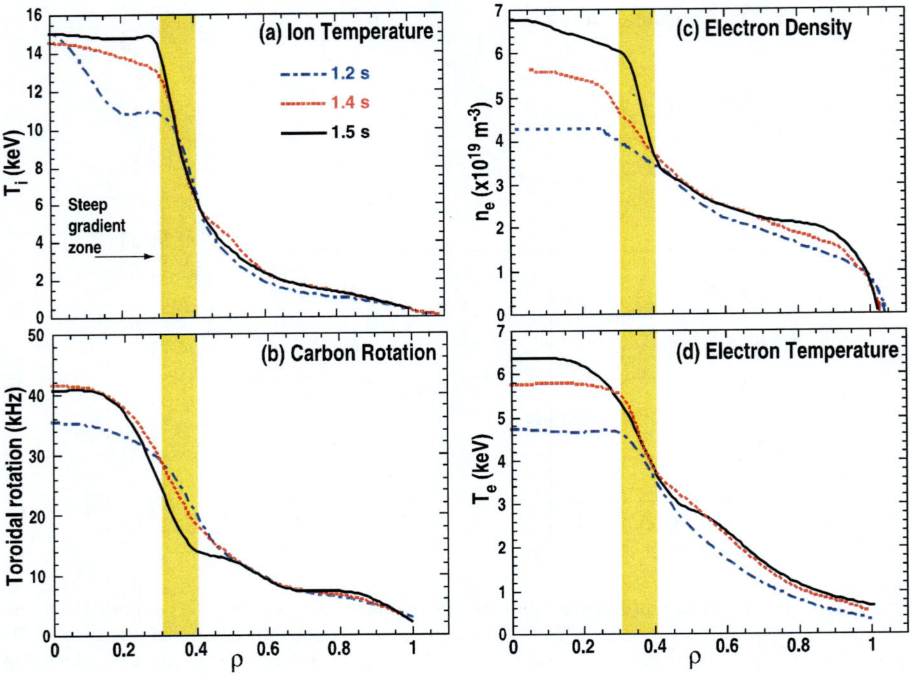

Fig. 5.25 Example of profile time evolution, illustrating differences in ITB development. Profiles are shown at 1.2 s (dashed-dot curve), 1.4 s (dashed curve) and 1.5 s (full curve) for (**a**) T_i, (**b**) carbon toroidal rotation, (**c**) n_e and (**d**) T_e for discharge 95,989. Figure 2 from [354] E.J. Doyle et al. 'Observation of simultaneous internal transport barriers in all four transport channels and correlation with turbulence behaviour in NCS discharges on DIII-D', Plasma Phys. Control. Fusion **42**, A237 (2000). ©IOP Publishing. Reproduced with permission. All rights reserved

theory [9, 358] of intrinsic rotation yields core rotation in H-mode in the co-current direction driven by the pressure gradient, also scaling inversely with B_θ. A related treatment of the residual stress, in which sheared radial electric fields break the symmetry $k_\parallel \rightarrow -k_\parallel$ of ITG mode turbulence [11, 326] reproduces the increase of intrinsic flow with $\nabla P/n$, and with a model of the pedestal width, recovers the $1/B_\theta$ scaling. Gyro-kinetic simulations confirm this model [359]. Expanding on this in the framework of a heat engine [8, 325, 328], treating the local fluctuation entropy balance in an ITG mode dominant regime, leads to Eq. 5.2, which is in quantitative agreement with the intrinsic rotation scaling observed in C-Mod plasmas (Sect. 5.1.2), replicating the direction, magnitude and scalings with ρ_*, ∇T and q_{95}. Similar intrinsic rotation production can occur due to collisionless trapped electron mode (CTEM) turbulence, where the drive is from electron temperature and density gradients. Global gyro-kinetic simulations of non-linear residual stress generation from zonal flow shear induced asymmetry in the $k_\parallel$ spectrum [360–362] reproduces the observed scalings with kinetic profile gradients and with an inverse plasma current dependence. A similar $\nabla T/I_p$ scaling is found from a treatment

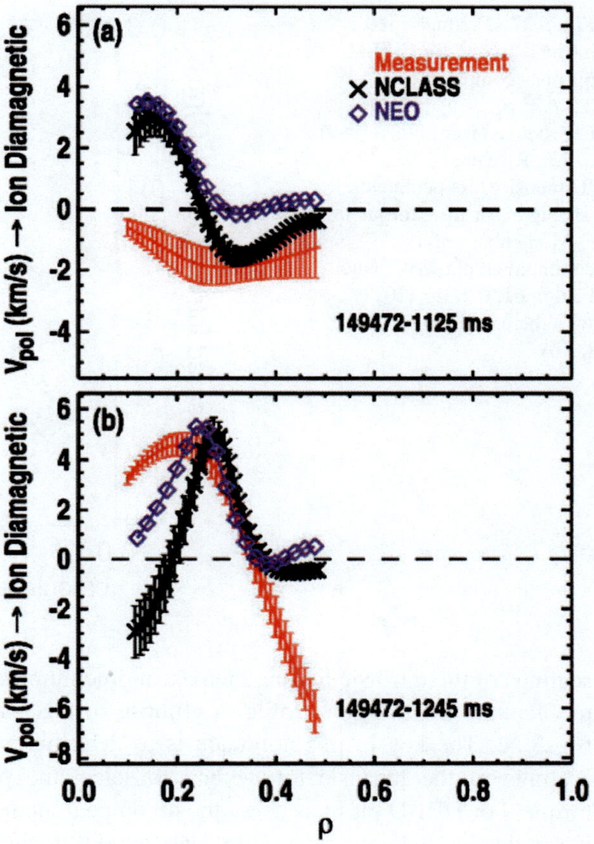

Fig. 5.26 Comparisons of measured and neo-classical poloidal rotation (**a**) before and (**b**) after ITB formation in DIII-D. Agreement is poor before and good after in the transport barrier region (0.2 $< \rho <$ 0.35). Reproduced from Fig. 7 of [293] C. Chrystal et al. 'Testing neoclassical and turbulent effects on poloidal rotation in the core of DIII-D', Phys. Plasmas **21**, 072504 (2014), with the permission of AIP Publishing

where the symmetry breaking is due to the turbulence intensity gradient [363]. Another source of intrinsic flow drive is from charge separation induced by the polarization drift [364]. A ∇P dependence of intrinsic rotation is also found from the polarization drift and its effect on trapped ions [365]. A mechanism for turbulent acceleration of intrinsic toroidal rotation due to the ion temperature gradient has been identified from gyro-kinetic theory [366]. Finally, considering higher order expansions in ρ_θ/L_T, a scaling is found for the intrinsic rotation $\propto T/I_p$ [210], similar to the results in Sect. 2.2.3 on edge orbit loss. This latter result is for hollow rotation profiles and matches experimental results in Ohmic L-mode plasmas and H-mode discharges with ECH. Consideration of the effects of diamagnetic flows in a similar higher order ρ_* expansion leads to an increase in intrinsic flow with ∇P [181]. Another approach which captures the ∇P rotation drive includes kinetic stresses in an electromagnetic gyro-kinetic treatment [367]. Discerning between these various theories is the challenge for understanding self generated flow in plasmas with transport barriers.

For the most part, the results of Sect. 5.1 have characterized intrinsic rotation in enhanced confinement regimes, which can then be used to predict self-generated flow in future devices, for example using Eq. 5.2. Another approach is to establish

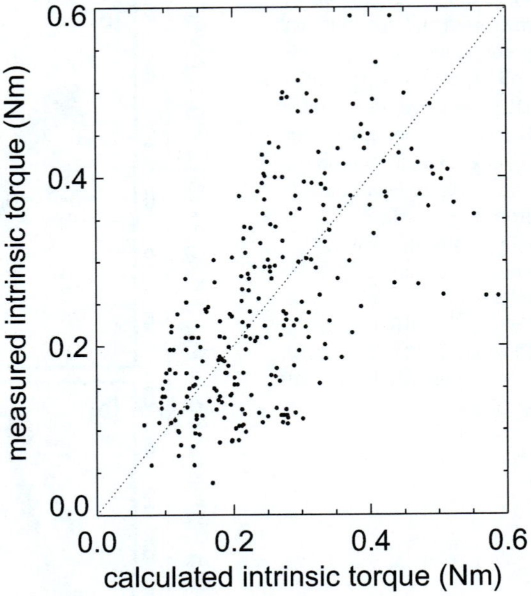

Fig. 5.27 The measured intrinsic torque for C-Mod plasmas compared to $T_{int} = 2.3\ \beta_N^{1.5}\ \rho_*^{-1.0}\ \nu_*^{0.1}$ Nm. Reproduced from Fig. 6 [369] of J.E. Rice et al. 'Dimensionless parameter scaling of intrinsic torque in C-Mod enhanced confinement plasmas', Nucl. Fusion **61**, 026013 (2021), with the permission of the IAEA

scalings of the intrinsic torque, then use momentum transport coefficients (Chap. 6) to calculate the rotation profile- a glimpse of this was covered in Sect. 5.1.5 and Fig. 5.22. The key to this approach is to determine the dimensionless parameter scalings of the intrinsic torque and to select the proper normalization for the torque. For DIII-D plasmas using the onion peel method [110], the intrinsic torque, normalized to the ion temperature, was found to scale as $\rho_*^{-1.5\pm0.8}\ \nu_*^{0.26\pm0.04}$ [345]. A similar approach, using NBI modulation, has been applied to other devices [116, 368]. The instrinsic torque has been determined in C-Mod plasmas by a simple method, comparing the total angular momentum of the plasma before and after L-H or L-I transitions, and dividing by the time for the profile to evolve. This approach has been applied to the same database of discharges shown in Fig. 5.11, and the best fit for the intrinsic torque, $T_{int} = 2.3\ \beta_N^{1.5}\ \rho_*^{-1.0}\ \nu_*^{0.1}$ Nm, is shown in Fig. 5.27 [369]. There is a strong dependence on β_N, and similar scalings with ρ_* and ν_* as was found for DIII-D discharges. Comparison of these results to those from other devices should provide information on the proper normalization for the intrinsic torque. The choice of the ion temperature for this purpose was not very satisfying. The best results were achieved normalizing the intrinsic torque to the device major radius, and the results are displayed in Fig. 5.28 [369]. These points were from discharges with matched β_N, ν_*, q_{95} and T_e/T_i. The curve allows for estimates to be made for future devices (with $\beta_N \sim 1.3$). For ITER, with R = 6.2 m and $\rho_* = 0.0015$, the intrinsic torque is estimated to be ∼20 Nm, which is somewhat close to the value of 33 Nm predicted in [345] using a normalization proportional to the ion temperature. For SPARC [370], with R = 1.85 m and an estimated $\rho_* = 0.0027$, Fig. 5.28 gives an intrinsic torque of ∼4 Nm, while for ARC [330], with R = 3.3 m and $\rho_* = 0.002$, T_{int} is expected to be ∼8 Nm.

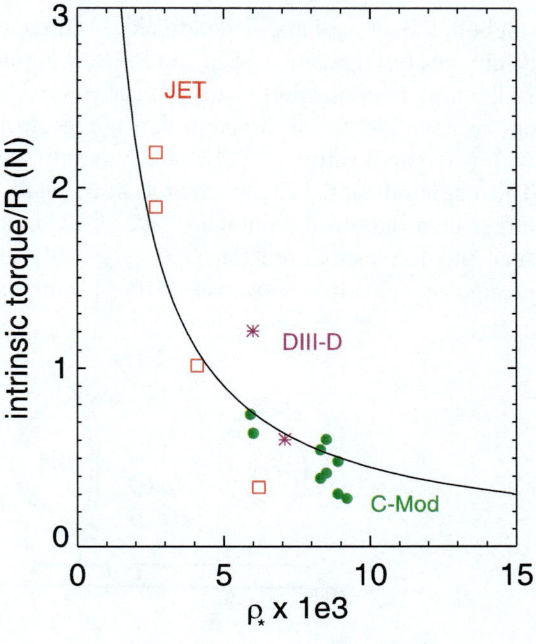

Fig. 5.28 The intrinsic torque normalized to the major radius as a function of ρ_* for JET (red squares), DIII-D (purple asterisks) and C-Mod (green dots) for plasmas with β_N near 1.28. The curve is proportional to $\rho_*^{-1.0}$. Reproduced from Fig. 9 [369] of J.E. Rice et al. 'Dimensionless parameter scaling of intrinsic torque in C-Mod enhanced confinement plasmas', Nucl. Fusion **61**, 026013 (2021), with the permission of the IAEA

5.2 Low Confinement Mode

Intrinsic toroidal rotation in Ohmic low confinement mode (L-mode) plasmas is more complicated and exhibits dependences on the density, plasma current, toroidal magnetic field, impurity content, central safety factor and magnetic configuration. In Sect. 5.2.1 the velocity response to changes in magnetic configuration will be shown. The effects of current drive are presented in Sect. 5.2.2. Ohmic rotation reversals will be summarized in Sect. 5.2.3 where the dependence on collisionality will be presented. Rotation reversals in auxiliary heated plasmas will be shown in Sect. 5.2.4.

5.2.1 Magnetic Configurations

By simply changing the X-point from lower single null (LSN) to double null (DN) to upper single null (USN), or to inner wall limited (IWL), the scrape-off-layer (SOL) flows can change drastically [74, 75, 371]. This is demonstrated in Fig. 5.29 which in the top two frames shows SOL velocities as a function of the parameter SSEP, in a shot by shot scan in C-Mod Ohmic L-mode plasmas with a density of 1.5×10^{20}/m^3, and the ion $\mathbf{B}\times\nabla\mathbf{B}$ drift downward [75]. SSEP is the distance between the primary and secondary separatrices, mapped to the outer midplane, and is a measure of the configuration state, as illustrated at the top of the figure. LSN plasmas have negative values of SSEP, and the SOL flow a couple of millimeters outside of the separatrix (on both the inboard and outboard side) is directed co-current. Conversely for USN discharges (with positive SSEP) the SOL flows are in the counter-current direction. Ballooning like transport leads to a net volume averaged SOL toroidal momentum, co- or counter-current directed, depending on the X-point location. DN plasmas, with very small values of SSEP, are smoothly connected to those in LSN and USN. The magnitude of the change in SOL flows on the inboard side was ~50 km/s, much larger than that on the outboard side, ~12 km/s. Qualitatively similar behavior is seen in the plasma core [36, 320, 372–374] (bottom frame of Fig. 5.29) with a change of ~20 km/s. However, at the plasma center, under these conditions, the

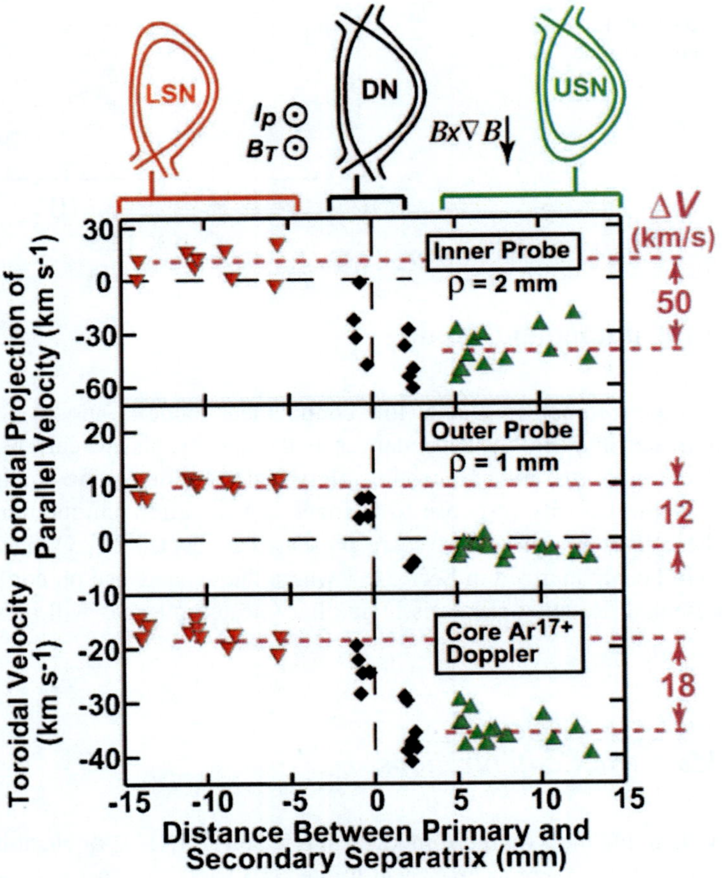

Fig. 5.29 Toroidal projection of parallel flows measured near the separatrix by probes (top two panels) and toroidal rotation velocity measured at the plasma center (bottom panel). Positive velocity is in the direction of plasma current. The distance between the primary and secondary separatrix surfaces (mapped to the outer midplane) is used to parameterize the upper/lower X-point balance. Reproduced from Fig. 4 of [75] B. LaBombard et al. 'Transport-driven scrape-off layer flows and the x-point dependence of the L-H power threshold in Alcator C-Mod', Phys. Plasmas **12**, 056111 (2005), with the permission of AIP Publishing

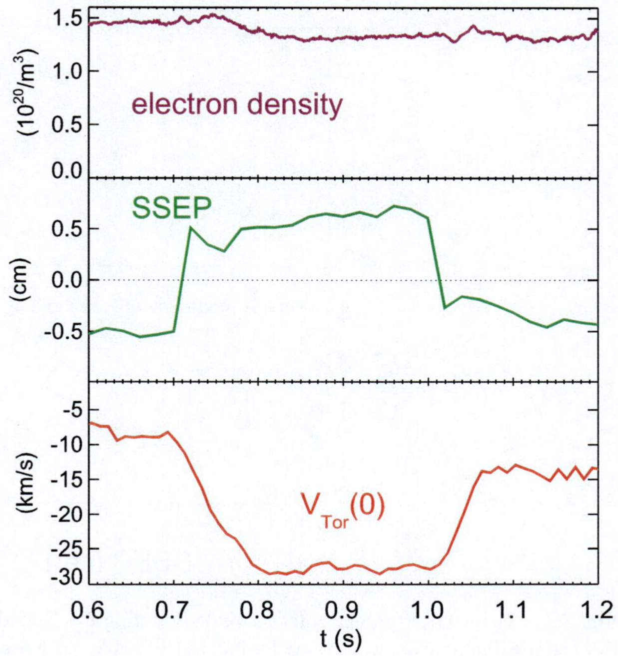

Fig. 5.30 Time histories of the electron density (top frame), SSEP (middle frame) and the central toroidal rotation velocity for a C-Mod L-mode discharge which underwent an abrupt change from LSN to USN at 0.7 s and from USN back to LSN at 1.0 s

rotation is all counter-current. (Core rotation with different plasma parameters, will be covered in detail in Sect. 5.2.3.) It is possible to vary SSEP dynamically during a single discharge, and the same behavior is apparent [36]. Shown in Fig. 5.30 is the central rotation response to a sudden change from LSN to USN and back in a C-Mod Ohmic L-mode discharge. The change from LSN to USN occured at 0.7 s, after which time the rotation went from around −10 km/s to about −30 km/s, and back again at 1.0 s with a change from USN to LSN. This technique of an edge perturbation giving rise to a core change in the velocity can be used to characterize the momentum diffusivity, and will be discussed in 6.1. The very strong counter-current rotation in limited discharges or plasmas with 'unfavorable' drift (ion $\mathbf{B} \times \nabla\mathbf{B}$ drift opposite to the X-point location) appears to be related to the higher H-mode power threshold observed in these configurations [75, 372]. This is demonstrated in Fig. 5.31 which shows both the core rotation and the H-mode power threshold as a function of SSEP for 5.4 T, 0.8 MA C-Mod discharges with $n_e = 1.4\times10^{20}/m^3$. The shapes of the trends are opposite, with the rotation velocity and threshold power being very sensitive to SSEP in near DN plasmas.

A systematic study of the dependence of toroidal rotation on the plasma shape has been performed on TCV [375]. An example of the rotation response to different shaping is shown in Fig. 5.32 for 340 kA, 1.4 T (q_{95} = 2.9) plasmas with $n_e = 4.1\times10^{19}/m^3$. Toroidal rotation velocity profiles are compared following changes in the up/down symmetry of the magnetic flux surfaces. These responses have been modeled through changes in Π^{res} due to symmetry breaking from the varying magnetic shear [376, 377]. Optimally tilting the triangularity can be used to drive

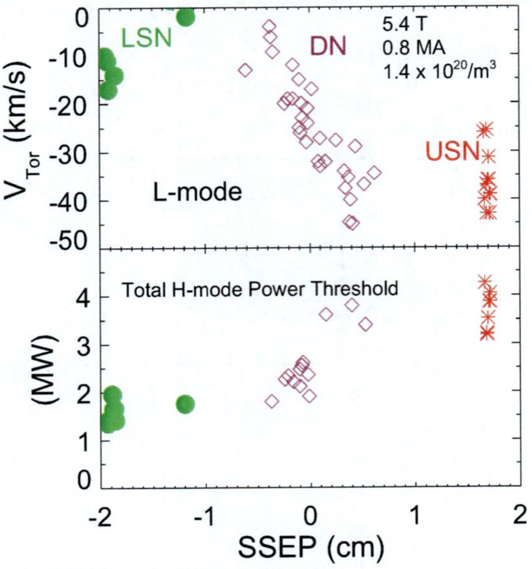

Fig. 5.31 The central toroidal rotation velocity during the L-mode portion of LSN (green dots), DN (purple diamonds) and USN (red asterisks), 0.8 MA, 5.4 T discharges with $n_e = 1.4 \times 10^{20}/m^3$ as a function SSEP is shown in the top frame. In the bottom is the minimum ICRF power required to induce the L-H transition as a function of SSEP, for several of these same discharges. Reproduced from Fig. 10 of [372] J.E. Rice et al. 'The dependence of core rotation on magnetic configuration and the relation to the H-mode power threshold in Alcator C-Mod plasmas with no momentum input', Nucl. Fusion **45**, 251 (2005), with the permission of the IAEA

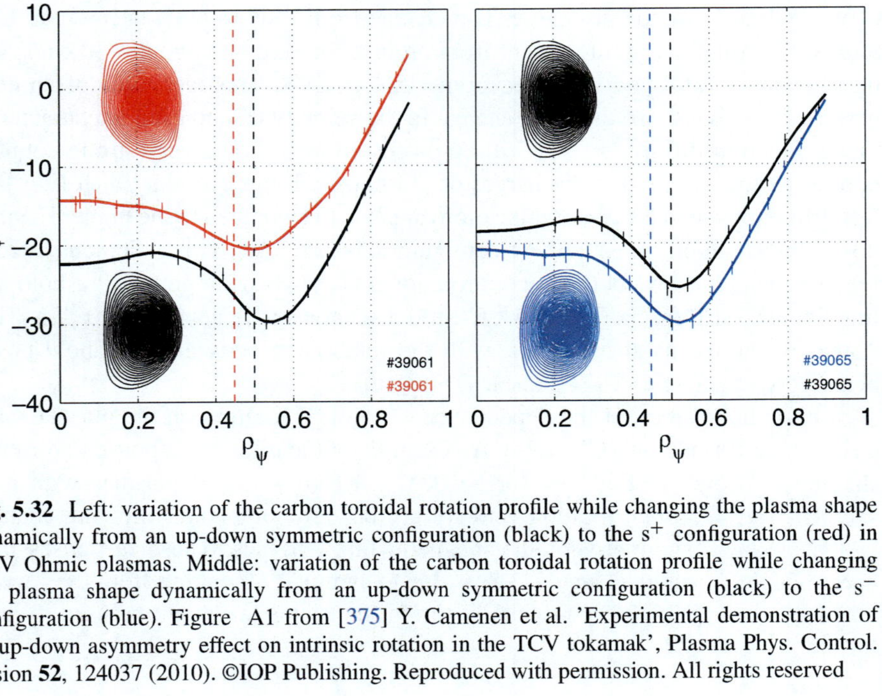

Fig. 5.32 Left: variation of the carbon toroidal rotation profile while changing the plasma shape dynamically from an up-down symmetric configuration (black) to the s^+ configuration (red) in TCV Ohmic plasmas. Middle: variation of the carbon toroidal rotation profile while changing the plasma shape dynamically from an up-down symmetric configuration (black) to the s^- configuration (blue). Figure A1 from [375] Y. Camenen et al. 'Experimental demonstration of an up-down asymmetry effect on intrinsic rotation in the TCV tokamak', Plasma Phys. Control. Fusion **52**, 124037 (2010). ©IOP Publishing. Reproduced with permission. All rights reserved

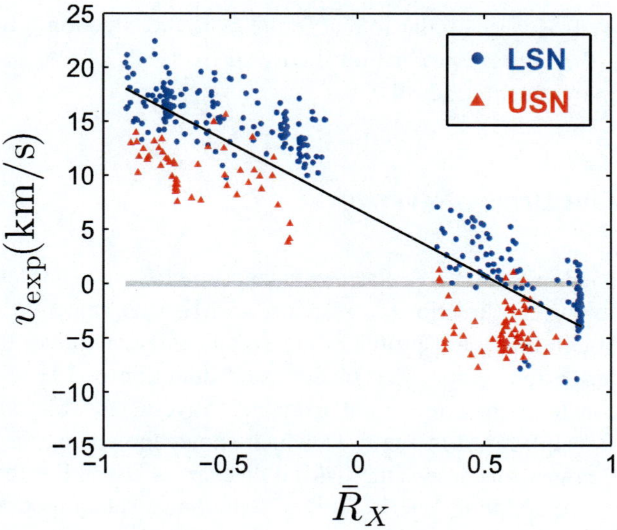

Fig. 5.33 Experimentally measured rotation at the core-edge boundary (v_{exp}) as a function of X-point position R_x. Measured rotation v_{exp} shifts strongly in the counter-current direction for increasing R_x, actually becoming counter-current for large enough R_x, consistent with the theoretical predictions. The slope and offset for the solid black fit line yield reasonable values for the two input parameters in Eq. 5.3, d_c and L_ϕ. Reprinted with permission from Fig. 4 of [379] T. Stoltzfus-Dueck et al. 'X-Point-Position-Dependent Intrinsic Toroidal Rotation in the Edge of the TCV Tokamak', Phys. Rev. Lett. **114**, 245001 (2015). Copyright (2015) by the American Physical Society

rotation [378]. Further experiments have been performed on TCV to investigate the effect of the X-point location on toroidal rotation [379, 380]. The X-point location has been systematically scanned from the HFS to the LFS, in both LSN and USN discharges. The toroidal rotation is found to be more counter-current with the X-point on the LFS, with both LSN and USN. The results of these experiments are summarized in Fig. 5.33 where the rotation velocity is shown as a function of the X-point position, R_x, for both LSN and USN plasmas [379]. These observations are in good agreement with a model of passing ion orbit shifts in the presence of a large edge turbulence intensity gradient [209, 381], which yields an expression for the main ion edge rotation

$$v = 0.104(d_c/2 - R_x^N)\frac{q}{L_\phi(\mathrm{cm})}\frac{T_i(\mathrm{eV})}{B_T(T)}(\mathrm{km/s}) \tag{5.3}$$

where q is the edge safety factor, L_ϕ is the e-folding length for radial decay of the turbulence intensity, T_i is the ion temperature at the boundary between the core and the edge, and B_T is the toroidal magnetic field. The parameter d_c captures the poloidal variation of the turbulence intensity, which approaches +2 in the strongly ballooning limit. Here R_x^N is the normalized X-point major radius, which varies from -1 at the innermost major radius to $+1$ at the outermost. Interestingly, this scaling with qT_i/B is very similar to that found from the edge ion orbit loss model of 2.2.3 [196], the entropy production theory of Sect. 5.1.2 [325] and the higher order expansion treatment in [210].

5.2.2 *Current Driven Reversals*

As was mentioned in Sect. 2.1.3, plasmas have been observed to rotate in both the counter- and co-current directions in response to LH wave injection. If the power level is low enough that the q profile is not significantly modified, the rotation is directed counter-current [166–173], in the usual direction of LH wave injection, and consistent with momentum input from the waves. This was apparent from Fig. 2.13 and is emphasized in Fig. 5.34 which shows the change in velocity as a function of LH power for these same C-Mod discharges [172]. For these discharge conditions (5.4 T, 0.8 MA, $n_e = 0.7 \times 10^{20}/m^3$), the linear relation between velocity

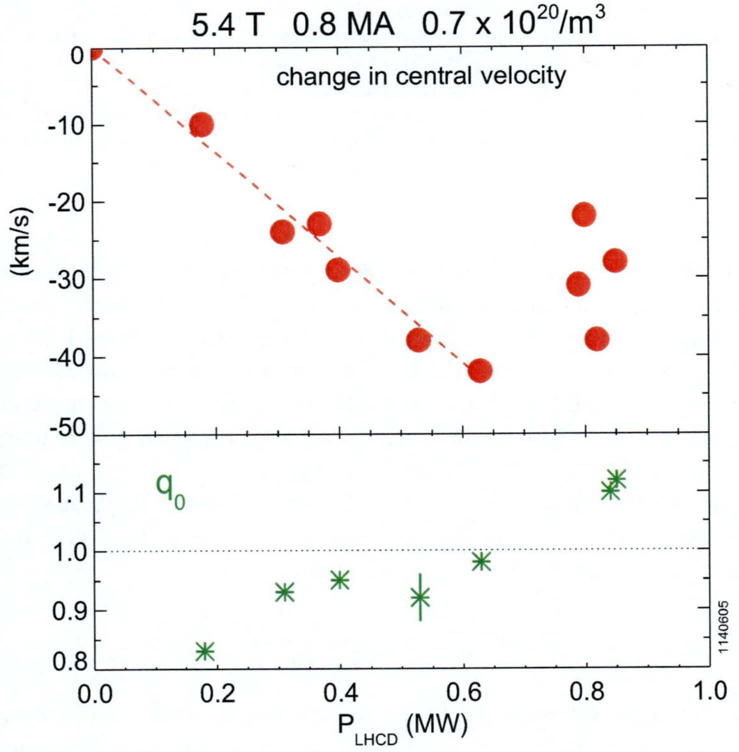

Fig. 5.34 Top frame: the change in the rotation velocity as a function of LHCD input power, evaluated at 1.45 s, for 0.8 MA discharges similar to those shown in Fig. 2.13. The dashed line indicates a linear fit to points below 0.65 MW. Bottom frame: the central inverse rotational transform, q_0, evaluated at 1.45 s. The horizontal line indicates $q_0 = 1$. Reproduced from Fig. 2 of [172] J.E. Rice et al. 'Effects of the q profile on toroidal rotation in Alcator C-Mod LHCD plasmas', Nucl. Fusion **56**, 036015 (2016), with the permission of the IAEA

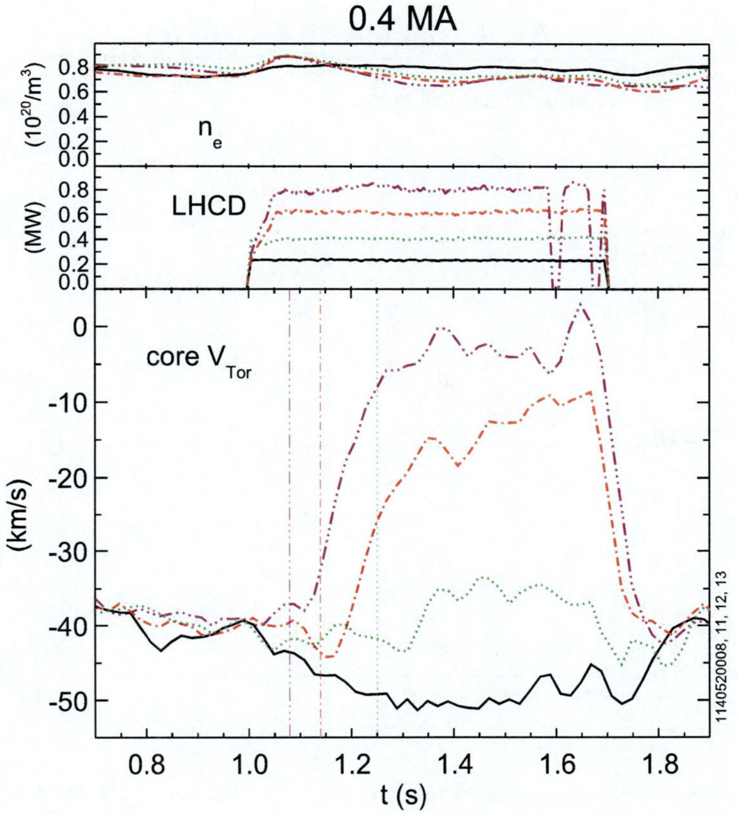

Fig. 5.35 Time histories of the electron density, LHCD power and core toroidal rotation velocity for 5.4 T, 0.4 MA discharges with different LHCD power levels. The vertical lines in the bottom frame indicate the times at which sawtooth oscillations disappeared for the three highest power discharges. Reproduced from Fig. 5 of [172] J.E. Rice et al. 'Effects of the q profile on toroidal rotation in Alcator C-Mod LHCD plasmas', Nucl. Fusion **56**, 036015 (2016), with the permission of the IAEA

change and LH power is broken above 0.7 MW; this coincides with enough current drive to bring q_0 above unity, as seen in the bottom frame. If the LH power is high enough to raise the central safety factor above 1 and suppress sawtooth oscillations, the rotation response is in the co-current direction, as has been seen on several devices [143, 166, 169–175, 177]. This point is further emphasized in a similar set of experiments at lower plasma current, where there is significantly more current drive for the same LH power. Shown in Fig. 5.35 are the parameter time histories for 0.4 MA C-Mod discharges at different LH power levels, to be compared with Fig. 2.13. At the lowest power level for this current, the change in the rotation is in the counter-current direction, similar to what was seen in Fig. 2.13. For higher LH power, the rotation change is in the co-current direction; these discharges all exhibited sawtooth suppression, indicating that q_0 was above unity. This point is further emphasized in Fig. 5.36 which in a similar fashion to Fig. 5.34, shows the

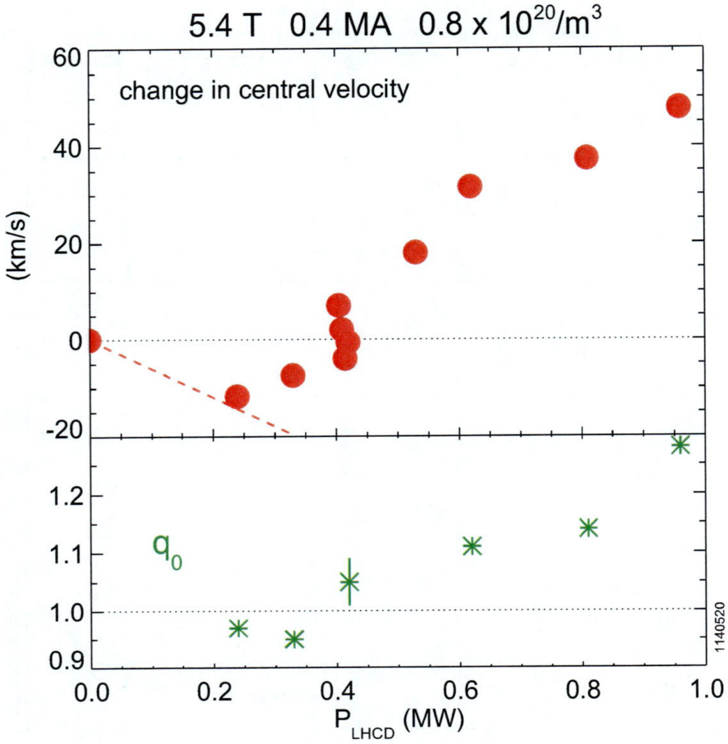

Fig. 5.36 Top frame: the change in the rotation velocity as a function of LHCD input power for the 0.4 MA plasmas of Fig. 5.33, evaluated at 1.6 s. The dashed line is the curve from Fig. 5.32, but with the slope reduced by the density ratio in the two scans, 88%. Bottom frame: The central inverse rotational transform, q_0, evaluated at 1.6 s. The horizontal line indicates $q_0 = 1$. Reproduced from Fig. 7 of [172] J.E. Rice et al. 'Effects of the q profile on toroidal rotation in Alcator C-Mod LHCD plasmas', Nucl. Fusion **56**, 036015 (2016), with the permission of the IAEA

change in velocity as a function of LH power for 0.4 MA C-Mod plasmas. For LH power above 400 kW, the rotation change is co-current, and this coincides with q_0 increasing above unity, and suppression of sawtooth oscillations. The results of Figs. 5.34 and 5.36 indicate a reconciliation of the co- and counter-current rotation observed in various LHCD experiments, and mentioned above. The rotation observations from several tokamaks are demonstrated in Fig. 5.37, which shows the change in rotation frequency in LHCD plasmas as a function of q_0 [170]. The data points occupy two quadrants, with counter-current directed rotation when the LH driven current is low enough so $q_0 < 1$ and co-current with $q_0 > 1$. The q_0 value for the JT-60U point is unknown, but was presumably above one since the sawteeth were suppressed. Several mechanisms have been put forward to explain both co- and counter-current driven rotation in LHCD plasmas (see Sect. 2.1.3) but the results summarized in Fig. 5.37 suggest that the current density profile is the primary ingredient. Changes in the residual stress through modifications of the q profile have recently been described [181–183].

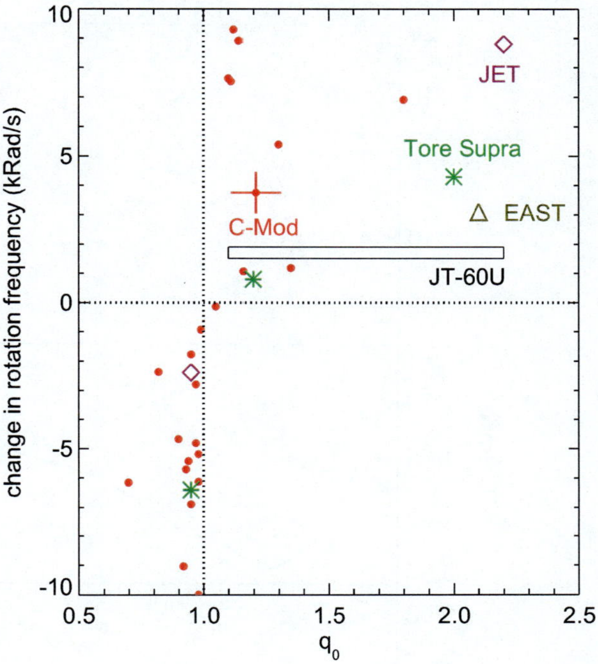

Fig. 5.37 The change in the core rotation frequency with LHCD as a function of q_0. Dots: C-Mod, asterisks: Tore Supra, diamonds: JET, triangle: EAST, box: JT-60U. Reprinted with permission from Fig. 6 of [170] J.E. Rice et al. 'Effects of Magnetic Shear on Toroidal Rotation in Tokamak Plasmas with Lower Hybrid Current Drive', Phys. Rev. Lett. **111**, 125003 (2013). Copyright (2013) by the American Physical Society

Modifications to the velocity profile due to changes in the q profile through ECH resonance location scans have been observed in DIII-D L-mode plasmas [333]. Rotation changes due to induced magnetic islands have been documented in J-TEXT discharges [250], and the effects of sawteeth on the velocity profile have been described for TCV plasmas [373], indicating other manifestations of q profile effects.

5.2.3 Ohmic Rotation Reversals and LOC/SOC

For the early velocity measurements in Ohmic L-mode plasmas [12, 14, 15, 17, 19], rotation in both the co- and counter-current directions was observed (see Table in [34]), giving the first hint of the rotation reversal phenomenon. Rotation reversals, changes in sign of the rotation, have been widely observed in Ohmic L-mode plasmas. Reversals were first observed through changes in the electron density, either dynamically via density ramps [19, 320, 373, 374, 382–393] or from shot by shot scans [320, 373, 374, 383, 385, 386, 389, 394–396]. An example of dynamic rotation reversals under the influence of density ramps down and then up is shown in Fig. 5.38 for a 1.1 MA, 5.4 T C-Mod discharge [384]. As can be seen by the solid line in the top frame, the core plasma experienced two reversals, first from the counter- to co-current direction begining at 0.666 s, following a slight decrease in the electron density (middle frame), and from co- to counter-current initiating at 1.146 s, with a slight density increase. There was no change in the rotation profile

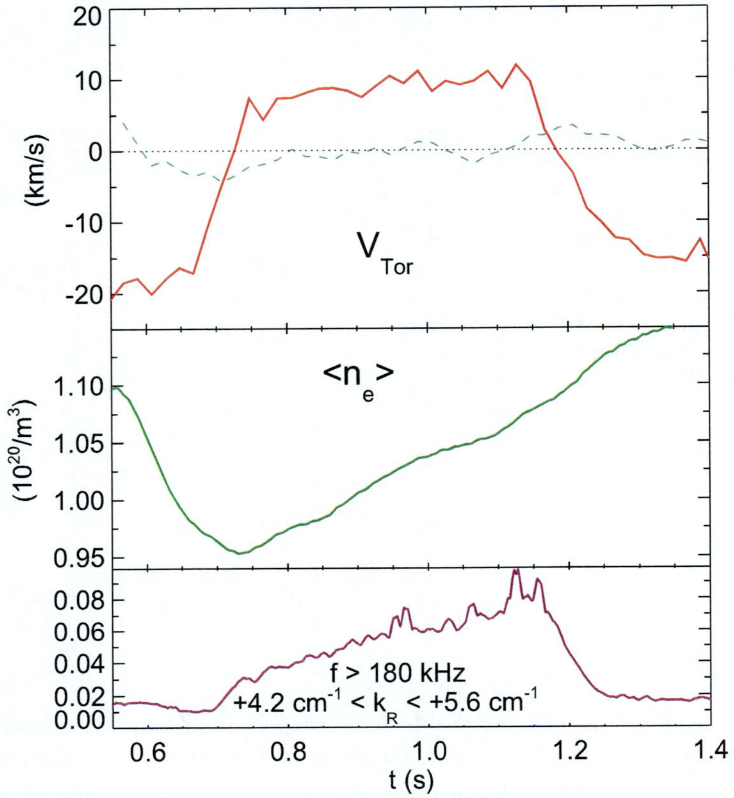

Fig. 5.38 Time histories of the toroidal rotation velocity (top frame) in the plasma center (solid line) and at r/a = 0.75 (dashed line), average electron density (middle frame) and core density fluctuation intensity (bottom frame) for a 1.1 MA, 5.4 T (q_{95} = 3.2) C-Mod discharge which underwent two rotation reversals. Reprinted with permission from Fig. 1 of [384] J.E. Rice et al. 'Rotation Reversal Bifurcation and Energy Confinement Saturation in Tokamak Ohmic L-Mode Plasmas', Phys. Rev. Lett. **107**, 265001 (2011). Copyright (2011) by the American Physical Society

outside of r/a = 0.7 during the core reversal, as can be seen by the dashed line in the top frame. Note that the scale for the density trace (middle frame) is greatly expanded, and the overall cycle occurred with a mere 10% change in the electron density. There was an increase in core density fluctuations (bottom frame) when the core plasma was spinning in the co-current direction. For a given density, the core can exist in two different rotation states. The existence of this hysteresis cycle [374, 382, 383] is *prima facie* evidence that the reversal is in fact a bifurcation and a first order transition. The hysteresis of the reversal process is exhibited in Fig. 5.39, which shows the discharge trajectory in the n_e-V_{Tor} plane [383]. As the density was lowered from 1.10 to $0.98\times10^{20}/m^3$ there was little change in the central rotation, which hovered ~ -20 km/s. Going below $0.98\times10^{20}/m^3$ caused the velocity to reverse direction and remain near $+10$ km/s until the density reached $1.07\times10^{20}/m^3$, at which time the rotation reversed again. The total torque density required to

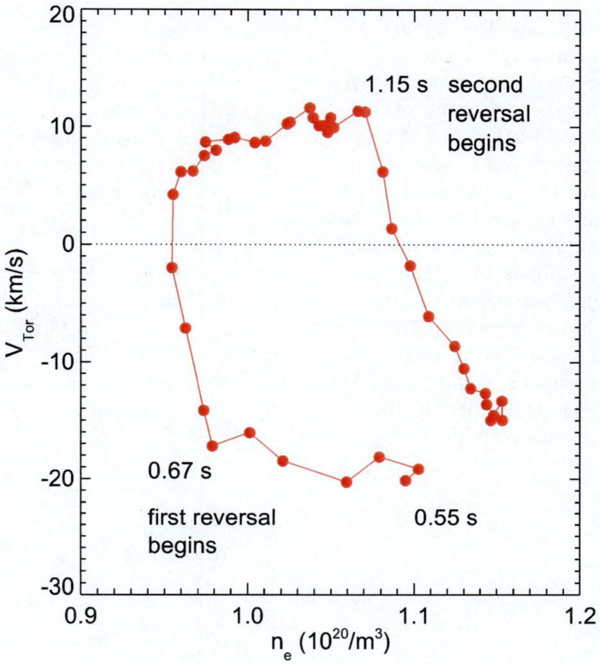

Fig. 5.39 The discharge trajectory in the n_e–V_{Tor} plane for the plasma of Fig. 5.38. Points are separated by 20 ms. Reproduced from Fig. 4 of [383] J.E. Rice et al. 'Observations of core toroidal rotation reversals in Alcator C-Mod ohmic L-mode plasmas', Nucl. Fusion **51**, 083005 (2011), with the permission of the IAEA

reverse the rotation in the core was ∼0.4 N/m^2. This is similar in magnitude to the measured Reynolds stress torque in DIII-D plasmas [76]. Rotation reversals at a critical density are apparent from shot to shot density scans, as can be seen in Fig. 5.40 for 5.4 T C-Mod discharges [386]. The critical reversal density increases with plasma current at fixed magnetic field [374]. For 1.1 MA plasmas at 5.4 T, the reversal density was $\sim 1.1\times10^{20}$/m^3, dropped to $\sim 0.8\times10^{20}$/m^3 for 0.8 MA and then to $\sim 0.6\times10^{20}$/m^3 for 0.55 MA. The critical reversal density for a wide range of plasma current is shown in Fig. 5.41 for 5.4 T C-Mod discharges [384].

The explicit relationship between rotation reversal and plasma current can be demonstrated through an I_p ramp as shown in Fig. 5.42 for a 6.3 T C-Mod discharge [385]. As the current was ramped down at fixed density, the core rotation changed sign. The dependence of toroidal rotation on plasma current is presented in Fig. 5.43 for a shot to shot I_p scan at fixed density and magnetic field for C-Mod discharges [385]. The velocity reversed direction at around 0.9 MA for this density, which is consistent with the density ramp results at fixed current [383]. This trend with plasma current in Ohmic L-mode plasmas [143, 372] is the inverse of the current dependence in H-mode, I-mode and other enhanced confinement regimes, where the toroidal rotation velocity was found to scale inversely with I_p, as was presented in Sect. 5.1.

Rotation reversals are also governed by the toroidal magnetic field; results for a magnetic field ramp are shown in Fig. 5.44 [383]. This is a comparison of two C-Mod LSN discharges at 0.8 MA and fixed density, one of which had a ramp down

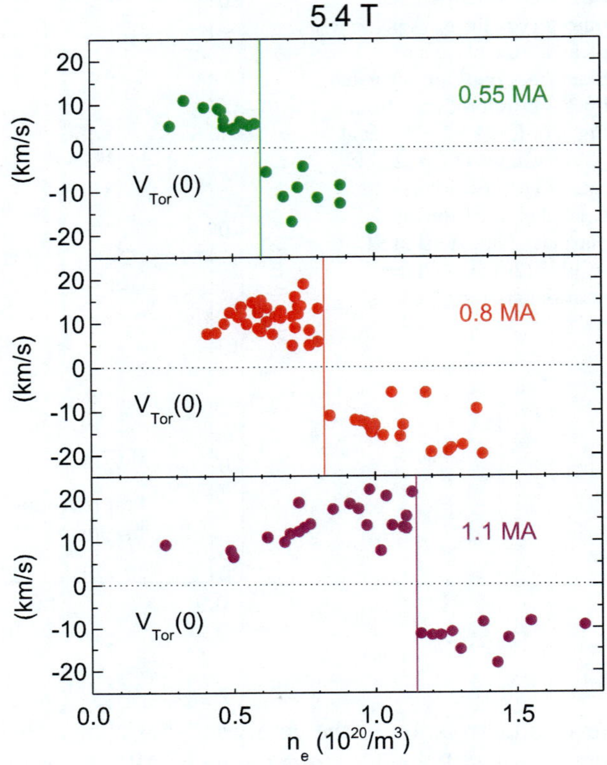

Fig. 5.40 The central rotation velocity as a function of average electron density, for plasma currents of 0.55 MA (top), 0.8 MA (center) and 1.1 MA (bottom). Adapted from Figs. 11, 12 and 13 from [386] J.E. Rice et al. 'Non-local heat transport, rotation reversals and up/down impurity density asymmetries in Alcator C-Mod ohmic L-mode plasmas', Nucl. Fusion **53**, 033004 (2013), with the permission of the IAEA

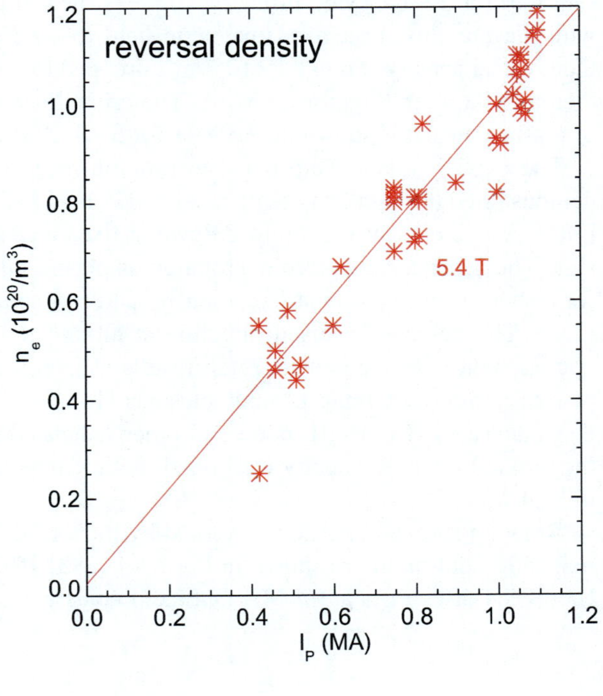

Fig. 5.41 The density at the rotation reversal initiation time as a function of plasma current for 5.4 T discharges. The line represents the best linear fit. Reprinted with permission from Fig. 2 of [384] J.E. Rice et al. 'Rotation Reversal Bifurcation and Energy Confinement Saturation in Tokamak Ohmic L-Mode Plasmas', Phys. Rev. Lett. **107**, 265001 (2011). Copyright (2011) by the American Physical Society

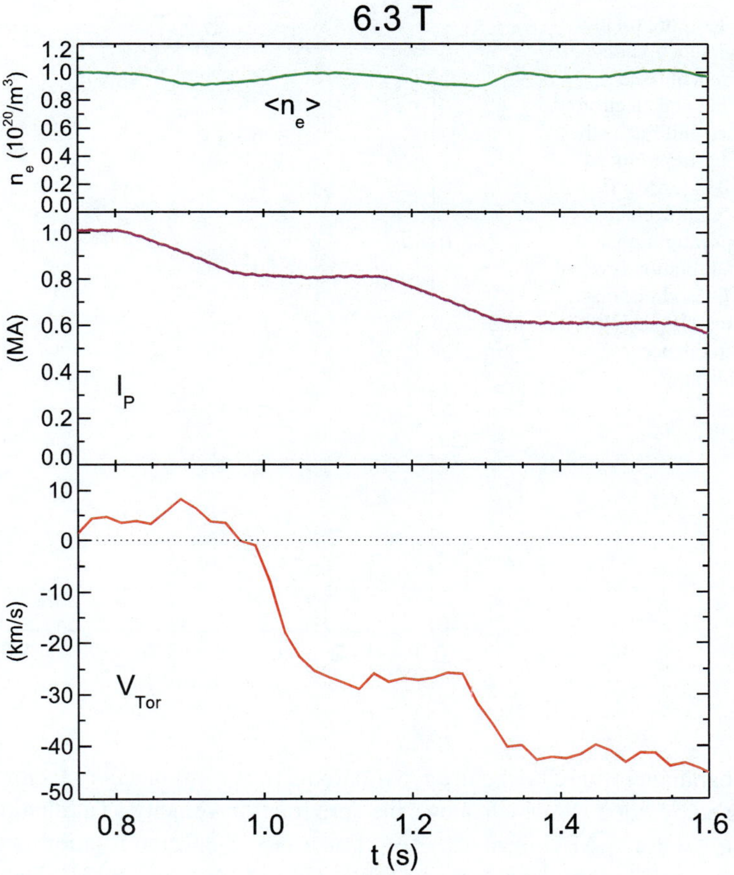

Fig. 5.42 The electron density (top), plasma current (middle) and core rotation velocity (bottom) for a 6.3 T discharge with downward density ramps. Reproduced from Fig. 9 of [385] J.E. Rice et al. 'Ohmic energy confinement saturation and core toroidal rotation reversal in Alcator C-Mod plasmas', Phys. Plasmas **19**, 056106 (2012), with the permission of AIP Publishing

and then up of the magnetic field. The plasma with the magnetic field ramp, from 4.5 to 3.0 T and back to 4.5 T, at a relatively constant density, experienced a rotation reversal during the ramp down, and a second reversal after the upward ramp. The discharge trajectory in the $q_{95} - V_{Tor}$ plane for the plasma with the magnetic field ramp (solid line in Fig. 5.44) is shown in Fig. 5.45. This figure demonstrates the hysteresis with magnetic field (the plasma current was held fixed) since for $q_{95} \sim 3$, there were both positive and negative values for rotation at this density. Shot by shot scans of the magnetic field at fixed current and density also reveal the rotation reversal dependence as shown in Fig. 5.46 [320]. For this density and current, the rotation reverses direction around 3.3 T. This behavior has been shown to be related to the dependence of the H-mode power threshold on magnetic field [320, 372]. The dependences of the rotation reversal phenomenon on I_p and B_T suggest that the

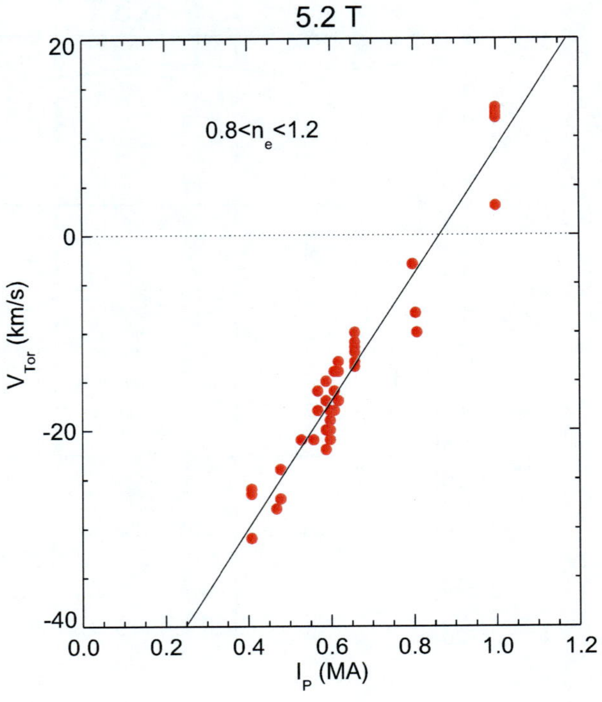

Fig. 5.43 The core rotation velocity as a function of plasma current at fixed magnetic field and electron density. The solid line is the best linear fit. Reproduced from Fig.10 of [385] J.E. Rice et al. 'Ohmic energy confinement saturation and core toroidal rotation reversal in Alcator C-Mod plasmas', Phys. Plasmas **19**, 056106 (2012), with the permission of AIP Publishing

relevant parameter might be the inverse rotational transform or safety factor, q. This is supported by Fig. 5.47 which shows the core rotation velocity as a function of the product $n_e q_{95}$ for C-Mod plasmas [169]. For $n_e q_{95} < 3.7$, the rotation is directed co-current, and abruptly switches to counter-current for $n_e q_{95} > 3.7$. Data points for discharges with parameters ranging from 0.3 to $2.0 \times 10^{20}/\text{m}^3$ for the density and from 2.6 to 7.2 for q_{95} are combined nicely compared to Fig. 5.40. This suggests that the relevant parameter might be the collisionality ν_*, which for fixed machine size, is proportional to $n_e q_{95}$.

Rotation reversal is not the only transition which occurs at a critical value of $n_e q_{95}$, and in order to understand the underlying mechanism for reversals, it is informative to consider these other phenomena. It is well known that there is a critical density for the transition from the linear Ohmic confinement (LOC) regime to the saturated Ohmic confinement (SOC) state (see the first 16 references in [385], which includes observations from at least 10 individual tokamaks). There is a large body of evidence which shows that the critical density for the LOC/SOC transition increases with plasma current [383, 385, 386, 395, 397–400]. The case that confinement saturation and rotation reversal are related [383–386, 395, 400–402] is supported by Fig. 5.48 which shows a comparison of core toroidal rotation velocity and energy confinement time in shot by shot density scans at fixed magnetic field (5.4 T) from C-Mod [386]. The vertical lines indicate the rotation reversal critical densities for each plasma current and they coincide nicely with the confinement saturation densities. The values for rotation reversal are very clear as the sign change

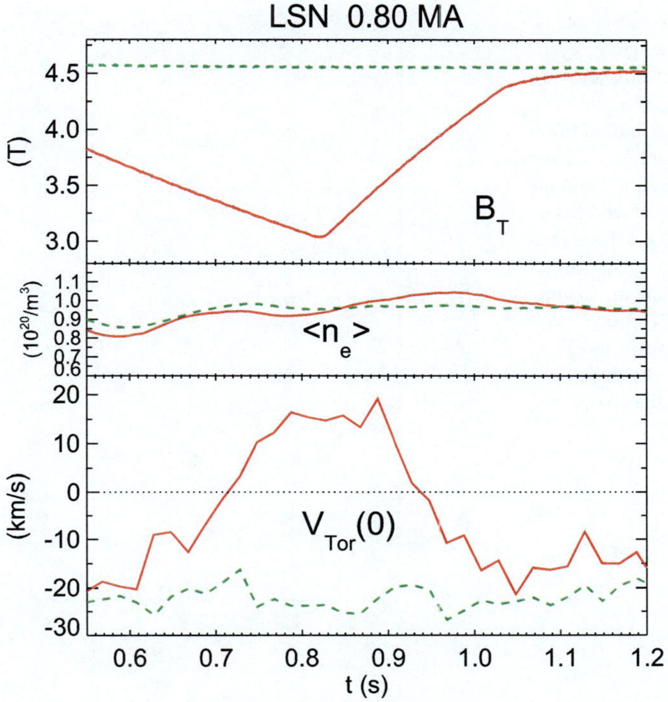

Fig. 5.44 Time histories of 0.8 MA LSN discharges with (solid) and without (dashed) a magnetic field ramp. From top to bottom, the toroidal magnetic field, the electron density and the central toroidal rotation velocity. Reproduced from Fig. 6 of [383] J.E.Rice et al. 'Observations of core toroidal rotation reversals in Alcator C-Mod ohmic L-mode plasmas', Nucl. Fusion **51**, 083005 (2011), with the permission of the IAEA

in the rotation is unequivical; on the other hand, the LOC/SOC transition density is harder to determine since it requires a large number of discharges with a long steady state for high quality kinetic profiles. This relationship is further demonstrated in Fig. 5.49 which compares the LOC/SOC (top) and rotation reversal (bottom) critical densities as a function of plasma current for C-Mod discharges at fixed B_T [385]. Both dotted lines have the same slope.

Another phenomenon that appears to be related is the so-called 'non-local' cutoff [386, 389, 403], which is demonstrated in the second panel of Fig. 5.50 from C-Mod. The 'non-local' cutoff (second frame), indicated by a reversal of the change in the mid-radius electron temperature following an abrupt drop in the edge temperature after a cooling event [386, 389], aligns well with the rotation reversal (third frame) and energy confinement time saturation (top frame). There also appears to be a peak in the inverse density scale length (bottom frame) at the same density. The relation among rotation reversal, the LOC/SOC transition and the 'non-local' cutoff is presented in Fig. 5.51, in the fashion of Fig. 5.49, except plotted against $1/q_{95}$

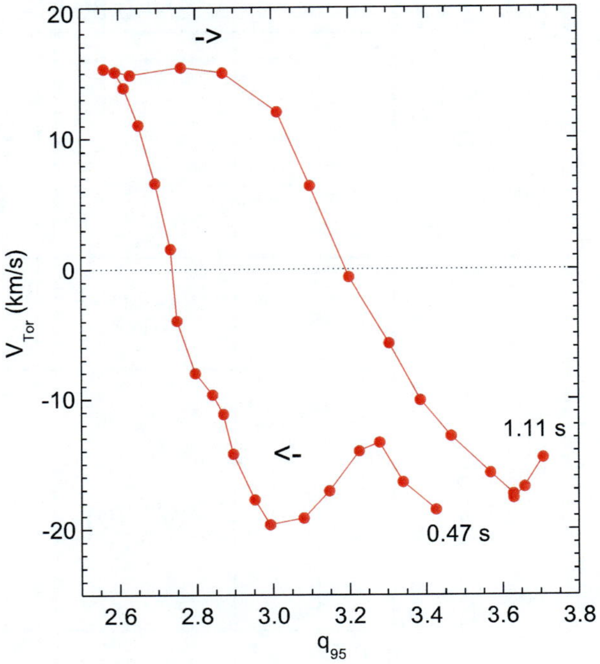

Fig. 5.45 The discharge trajectory in the $q_{95} - V_{Tor}$ plane for the plasma shown by the solid lines in Fig. 5.42 with the magnetic field ramp. Points are separated by 20 ms. Arrows indicate the temporal direction. Reproduced from Fig. 7 of [383] J.E. Rice et al. 'Observations of core toroidal rotation reversals in Alcator C-Mod ohmic L-mode plasmas', Nucl. Fusion **51**, 083005 (2011), with the permission of the IAEA

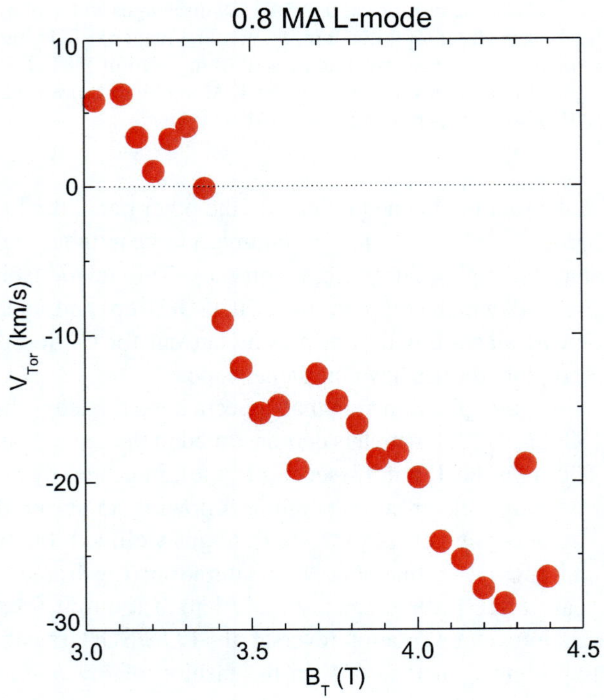

Fig. 5.46 The central toroidal rotation velocity as a function of the toroidal magnetic field for C-Mod 0.8, LSN discharges with an electron density of $1.0\times10^{20}/m^3$. Figure 4 from [320] J.E. Rice et al. 'Spontaneous core toroidal rotation in Alcator C-Mod L-mode, H-mode and ITB plasmas', Plasma Phys. Control. Fusion **50**, 124042 (2008). ©IOP Publishing. Reproduced with permission. All rights reserved.

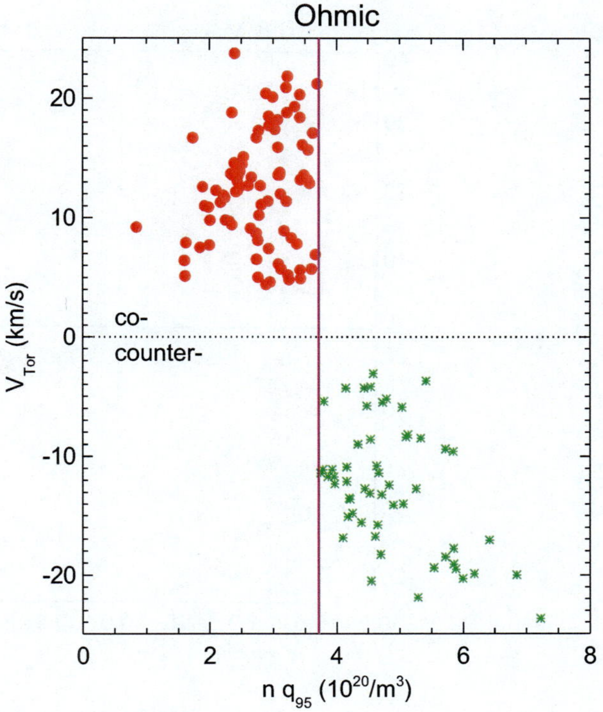

Fig. 5.47 The toroidal rotation velocity as a function of the nq_{95} product for Ohmic plasmas. The vertical line is $n_e q_{95} = 3.7$ with n_e in $10^{20}/m^3$. Reproduced from Fig. 11 of [169] J.E. Rice et al. 'Effects of LHRF on toroidal rotation in Alcator C-Mod plasmas', Nucl. Fusion **53**, 093015 (2013), with the permission of the IAEA

(which at fixed magnetic field is proportional to I_p). The trends are very similar although the points diverge a little at the highest current.

For confinement saturation, it is possible to determine the dependence on machine size taking advantage of observations from 13 different tokamaks. Shown in Fig. 5.52 is the density at the LOC/SOC transition at fixed q_{95} as a function of machine major radius, R [385]. These results are consistent with a 1/R dependence. Having established that rotation reversal and Ohmic confinement saturation have the same scaling, it is possible to explore the dependence of rotation reversal on collisionality, ν_*, the ratio of the collision frequency to the bounce frequency, $\approx 0.0118\, qRZ_{eff}n_e/\epsilon^{1.5}T_e^2$ with R in m, n_e in $10^{20}/m^3$ and T_e in keV. From Figs. 5.47 and 5.52, the critical density occurs for the product nRq = constant. For C-Mod Ohmic plasmas, the ratio Z_{eff}/T_e^2 is independent of density [385] and is independent of q_{95} at the reversal density, as can be seen in Fig. 5.53. The ratio Z_{eff}/T_e^2 and the critical ν_* for rotation reversal are both independent of plasma current. Rotation reversals occur at a fixed ν_*, equal to $\sim$0.35 at the reversal radius [385, 386]. This is the same value as was found on AUG [80] and MAST [390]. A model that includes the effect of small deviations from a Maxwellian equilibrium on turbulent momentum transport produces a rotation reversal for $\nu_* \sim \epsilon^{3/2}$ [181, 404], which captures these results.

In the previous discussion, the behavior of the central rotation velocity was documented. Further important information about the reversal process is revealed

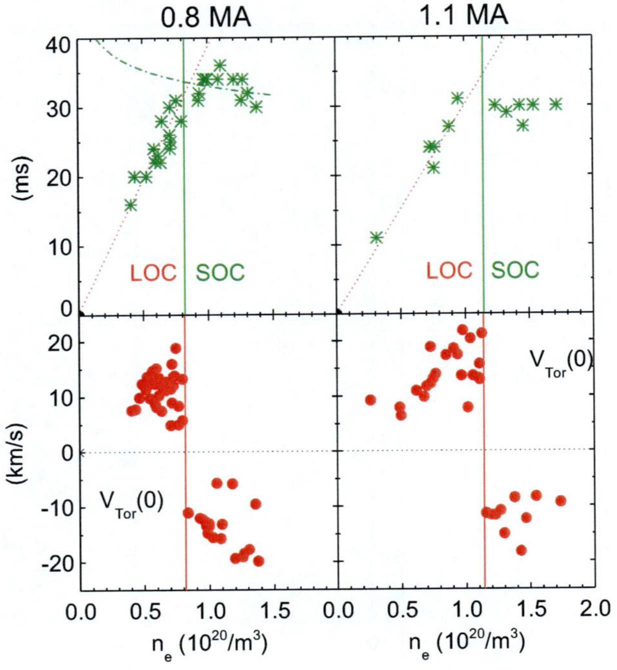

Fig. 5.48 Top: energy confinement times for 0.8 MA (left) and 1.1 MA (right) density scans at fixed magnetic field. Bottom: corresponding core rotation velocities. Vertical lines are from the rotation reversal critical densities. For energy confinement times, red dotted lines indicate neo-Alcator scaling in the LOC regime, and for the 0.8 MA case, the green dash-dot line represents the ITER-89P L-mode scaling. Adapted from Figs. 11 and 13 from [386] J.E. Rice et al. 'Non-local heat transport, rotation reversals and up/down impurity density asymmetries in Alcator C-Mod ohmic L-mode plasmas', Nucl. Fusion **53**, 033004 (2013), with the permission of the IAEA

through examination of the velocity profile shape. Figure 5.54 presents toroidal velocity spatial profiles on either side of a rotation reversal induced by an upward density ramp in a 5.4 T, 0.8 MA C-Mod discharge [386]. The profile in the LOC regime is flat across the minor radius while the shape with SOC is markedly hollow. Furthermore, the reversal occurs only inside of the anchor point, with very little change outside, as was seen in Fig. 5.38. These features are fairly generic and widely observed [295, 373, 382, 383, 386, 388, 391, 401, 405]. It is interesting that the rotation reversal anchor point coincides with electron temperature profile flex point for the 'non-local' behavior [386]. This location falls between the $q = 1$ and $q = 3/2$ radii. The position of the rotation reversal anchor point is a function of the plasma current, moving outward with increasing I_p [383, 386, 395]. This is demonstrated in Fig. 5.55 which shows the location of the rotation reversal anchor point as a function of $1/q_{95}$ ($\propto I_p$ at fixed B_T) for 5.4 T C-Mod discharges [386]. The anchor point moves outward with increasing plasma current and coincides with the 'non-local' T_e profile flex point. These fall between the calculated q=1 and q=3/2 surfaces. The

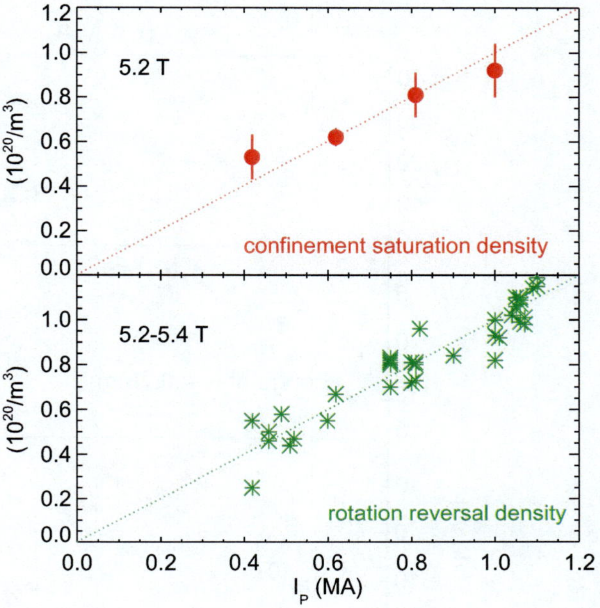

Fig. 5.49 The transition density between the LOC and SOC regimes (top) and the critical density for core toroidal rotation reversals (bottom) as a function of plasma current for fixed magnetic field. The dotted lines have the same slope. Reproduced from Fig. 3 of [385] J.E. Rice et al. 'Ohmic energy confinement saturation and core toroidal rotation reversal in Alcator C-Mod plasmas', Phys. Plasmas **19**, 056106 (2012), with the permission of AIP Publishing

steep gradient region of the velocity profile has been studied in detail [388, 391] and is found to be correlated with the electron density gradient.

It is important to keep in mind that most of the preceding C-Mod results (Figs. 5.40, 5.41, 5.48–5.51 and 5.53) were obtained near 5.4 T, and that combining observations from other tokamaks may reveal information about magnetic field dependences. The critical density for the LOC/SOC transition, rotation reversals and thermal non-diffusive cut-off have been combined in Fig. 5.56, a plot of the product of n_{crit}, R and the edge q value, as a function of toroidal magnetic field, for 18 individual tokamaks [406]. This contains all the points from Fig. 5.52, expanded to include all published edge q values. The LOC/SOC points represent contributions from 16 individual devices, and the rotation reversal and 'non-local' cut-off points are from 5 and 4 tokamaks each, respectively. The $n_{crit}qR$ product scales linearly with the magnitude of the magnetic field and probably explains some of the scatter apparent in Fig. 5.52. Examination of Fig. 5.56 indicates that the critical behavior occurs for nqR/B $\sim$ 0.5. nqR/B is very similar to n/n_G, and the suggestion is that there might be something interesting transpiring for $n/n_G \sim 1/2$. Figure 5.56 can be used for predicting the critical density for the LOC/SOC transition in future devices. For ITER operation at 5.3 T, the nqR product is expected to be around 2.5, which yields a critical density of $0.1\times10^{20}/m^3$ with R = 6.2 m and q_{95} = 3.5. For ARC (R = 3.3 m) [330] at 9.2 T and q_{95} = 3.5, the expected density is $0.3\times10^{20}/m^3$, whereas for SPARC [370] at 12 T, the product from Fig. 5.56 is about 5, and with R = 1.85 m and q_{95} = 3.5, the predicted density is $0.8\times10^{20}/m^3$. For JT-60SA [407], with B_T = 2.25 T, R = 2.96 m and q_{95} = 3.5, the critical density is estimated to be $\sim0.1\times10^{20}/m^3$ and for DTT [408] (B_T = 6 T, R = 2.2 m and q_{95} = 3.5) the value

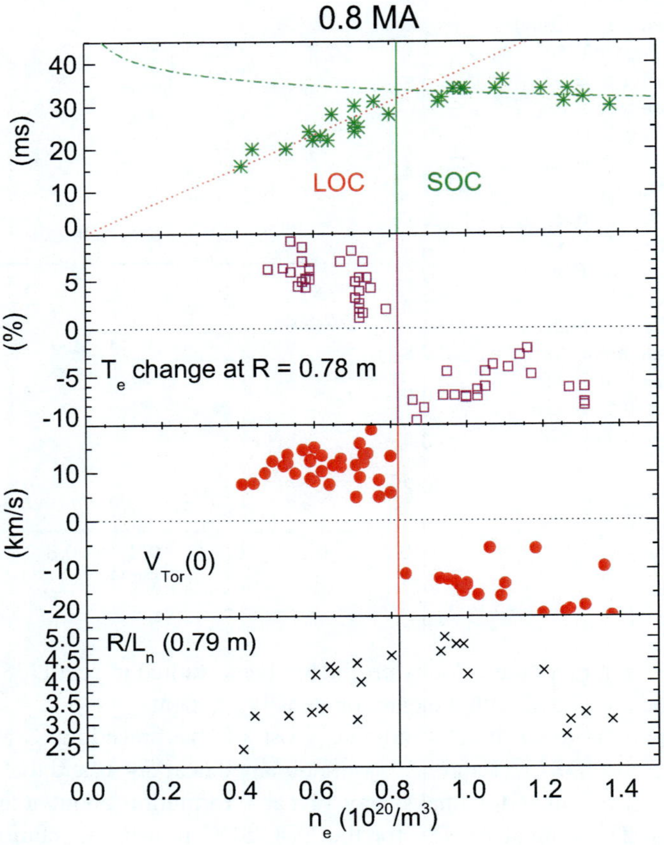

Fig. 5.50 The global energy confinement time (top), percent change in electron temperature at R = 0.78 m following impurity injections (second frame), core toroidal rotation velocity (third frame) and the inverse density gradient scale length (fourth frame) as a function of electron density for 0.8 MA, 5.4 T discharges. Reproduced from Fig. 11 of [386] J.E. Rice et al. 'Non-local heat transport, rotation reversals and up/down impurity density asymmetries in Alcator C-Mod ohmic L-mode plasmas', Nucl. Fusion **53**, 033004 (2013), with the permission of the IAEA

is $\sim 0.4 \times 10^{20}/\text{m}^3$. As a note, if reactors are chosen to operate in H-mode, the target plasmas will be in the SOC regime, otherwise the power threshold will be too high.

Further information about rotation reversals can be gleaned from inspection of the process dynamics. The time scale of the transition is of order the momentum confinement time [320, 373, 374, 382–384, 386, 390], 10 s of ms for C-Mod. A natural question arises concerning momentum conservation during the reversal process [382, 383], since the difference between the profiles with LOC and SOC is only apparent inside of the anchor point. Examination of the edge rotation with good time resolution reveals that during the core reversal process, there is a pulse of momentum which moves outward with the opposite sign to the core. This is documented in Fig. 5.57 which shows the dynamics of the reversal process in the

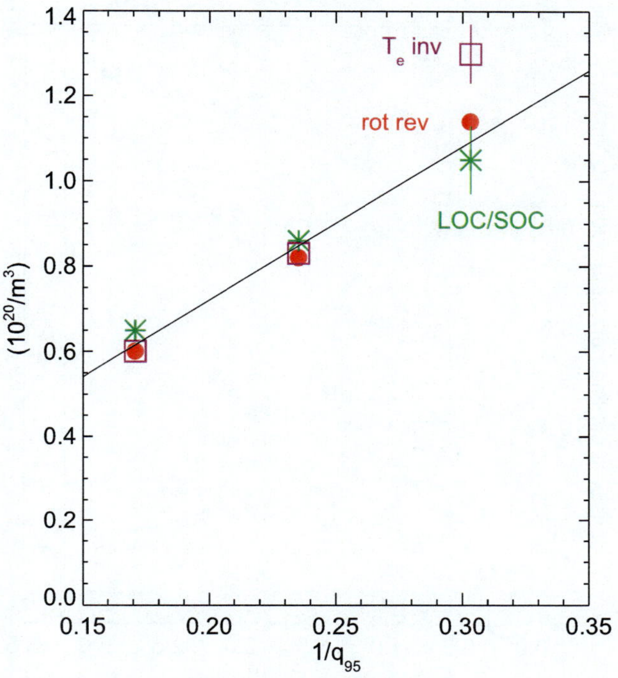

Fig. 5.51 The critical density for rotation reversal (dots), electron temperature inversion (boxes) and LOC/SOC transition (asterisks) as a function $1/q_{95}$. The line passes through $1/q_{95} = 0$. Reproduced from Fig. 15 of [386] J.E. Rice et al. 'Non-local heat transport, rotation reversals and up/down impurity density asymmetries in Alcator C-Mod ohmic L-mode plasmas', Nucl. Fusion **53**, 033004 (2013), with the permission of the IAEA

plasma center and near the edge for a C-Mod discharge with a density ramp [383]. During the reversal at the plasma center from counter- to co-current (solid black line), there is a transient pulse of momentum in the opposite direction at the plasma edge (red dots).

Another piece of the puzzle for the understanding of rotation revealed from inspection of turbulence changes across the LOC/SOC transition. Turbulence characteristics have been long known to change with confinement saturation [80, 121, 383–386, 389, 391, 394, 399, 400, 402, 409–414]. An example of the change in density fluctuations in the transition from LOC to SOC is demonstrated in Fig. 5.58 which shows dispersion plots S(k,f) of turbulence in the SOC regime with counter-current rotation (left) and with LOC and co-current flow (right). In the LOC regime there are distinct features at high k and high frequency which are absent with SOC. The time evolution of the intensity of this feature was shown in Fig. 5.38, which verifies that it is tied to the co-current rotation state with LOC. Changes in electron temperature fluctuations across the LOC/SOC transition have also been observed [412], as is shown in Fig. 5.59 which is a comparison of the coherence of T_e fluctuations in LOC (left) and SOC (right) plasmas from C-Mod. There is

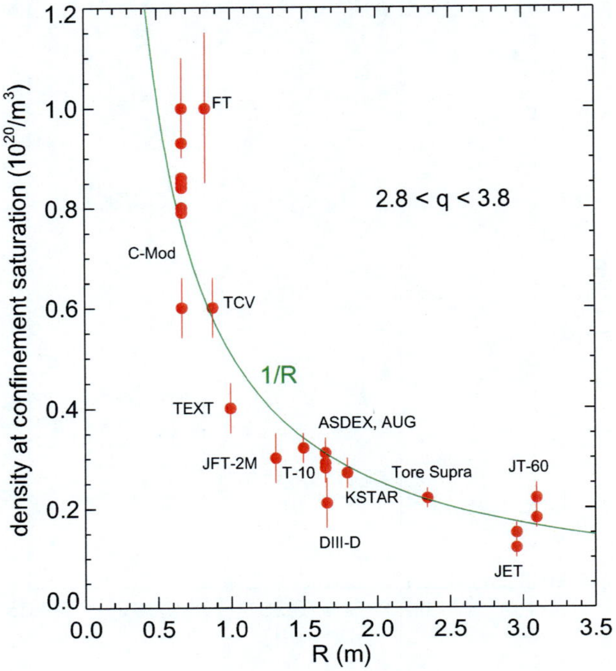

Fig. 5.52 The transition density from LOC to SOC as a function of major radius for different devices at fixed values of q. The solid curve represents 1/R. Reproduced from Fig. 7 of [385] J.E. Rice et al. 'Ohmic energy confinement saturation and core toroidal rotation reversal in Alcator C-Mod plasmas', Phys. Plasmas **19**, 056106 (2012), with the permission of AIP Publishing

Fig. 5.53 ν_* evaluated at the reversal radius (dots) and the ratio Z_{eff}/T_e^2 (asterisks) at the critical density for rotation reversal as a function of $1/q_{95}$. Reproduced from Fig. 21 of [386] J.E. Rice et al. 'Non-local heat transport, rotation reversals and up/down impurity density asymmetries in Alcator C-Mod ohmic L-mode plasmas', Nucl. Fusion **53**, 033004 (2013), with the permission of the IAEA

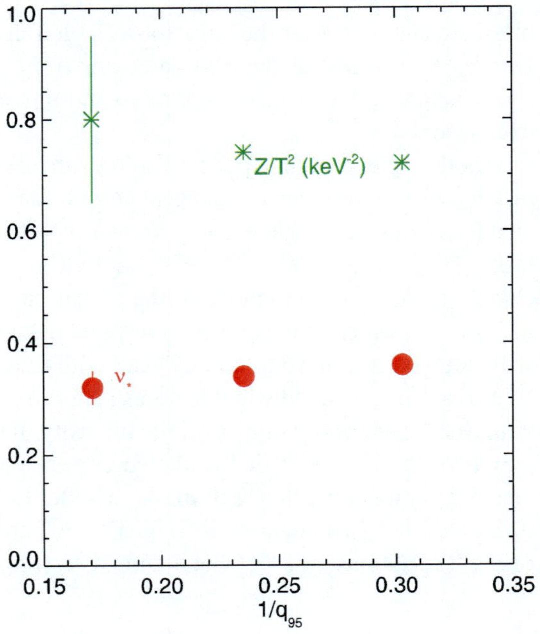

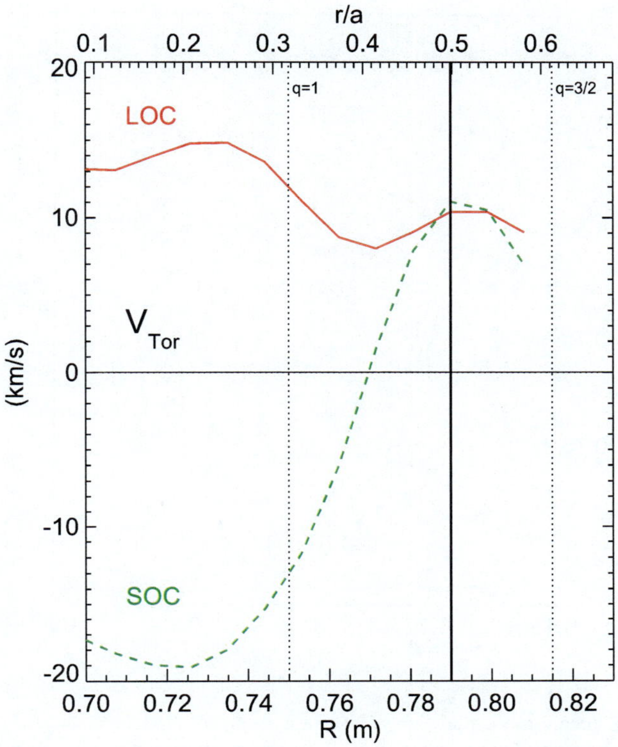

Fig. 5.54 Rotation velocity profiles at two different times for a discharge with a dynamic rotation reversal. The solid vertical line shows the rotation velocity profile anchor point. The dotted vertical lines indicate the locations of the q = 1 and q = 3/2 surfaces. Reproduced from Fig. 7 of [386] J.E. Rice et al. 'Non-local heat transport, rotation reversals and up/down impurity density asymmetries in Alcator C-Mod ohmic L-mode plasmas', Nucl. Fusion **53**, 033004 (2013), with the permission of the IAEA

a strong feature at ~50 kHz which is present with LOC and disappears following the transition to the SOC regime. Similar changes in microwave spectra have been widely observed across the LOC/SOC transition [80, 383, 391, 394, 399]. Examples from Tore Supra are shown in Fig. 5.60 which is a comparison of frequency spectra (and energy confinement times) at various densities for two different plasma currents [399]. 'QC modes' at ~60 kHz are apparent in LOC plasmas, which disappear after the confinement saturation. The dependence of the critical density for the LOC/SOC transition on plasma current, similar to that presented in Fig. 5.48, is cleary shown at the top. Tying together the behavior just documented suggests that in LOC plasmas with co-current rotation, certain manifestations of a turbulent state are visible, which disappear at a critical collisionality. Indeed, a change in sign of intrinsic rotation transpires with a change in sign of Π^{res}, which can occur if the propagation direction of the underlying turbulence switches from electron dominated to ion dominated. It has long been speculated that in the low collisionality LOC regime, the effects

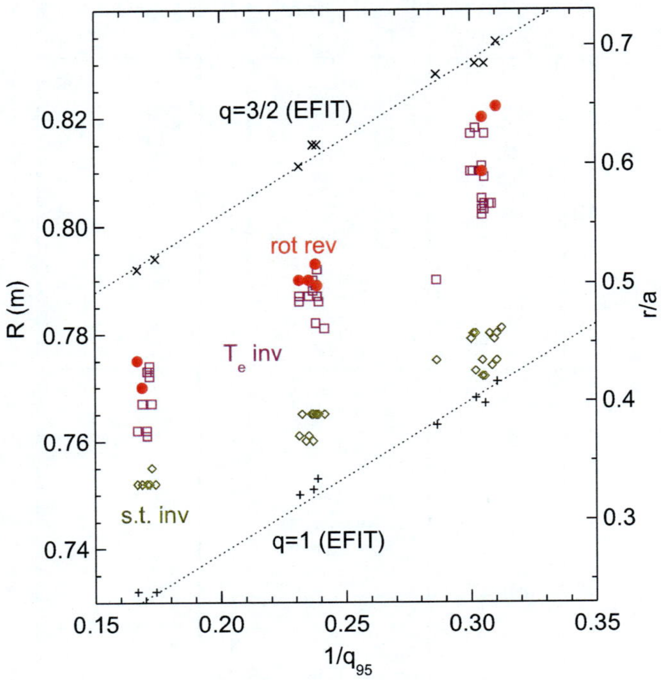

Fig. 5.55 Radial locations of the q = 3/2 surface (×s), rotation reversal anchor point (dots), electron temperature profile flex point (boxes), sawtooth inversion radius (diamonds) and q = 1 surface (plus signs) as a function of $1/q_{95}$. Reproduced from Fig. 16 of [386] J.E. Rice et al. 'Non-local heat transport, rotation reversals and up/down impurity density asymmetries in Alcator C-Mod ohmic L-mode plasmas', Nucl. Fusion **53**, 033004 (2013), with the permission of the IAEA

Fig. 5.56 The product of n_{crit}, the edge q value and R as a function of toroidal magnetic field from 18 individual tokamaks. Red dots are from the LOC/SOC transition, green asterisks are from rotation reversals and purple triangles are from the "non-local" cut-off. Reproduced from Fig. 14 of [406], J.E. Rice et al. 'Understanding LOC/SOC phenomenology in tokamaks', Nucl. Fusion **60**, 105001 (2020), with the permission of the IAEA

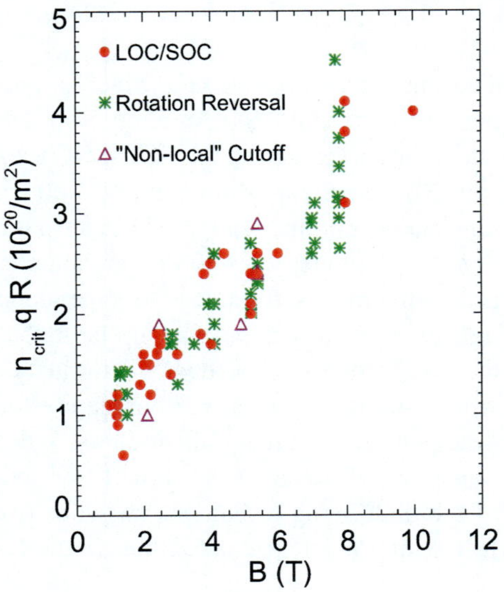

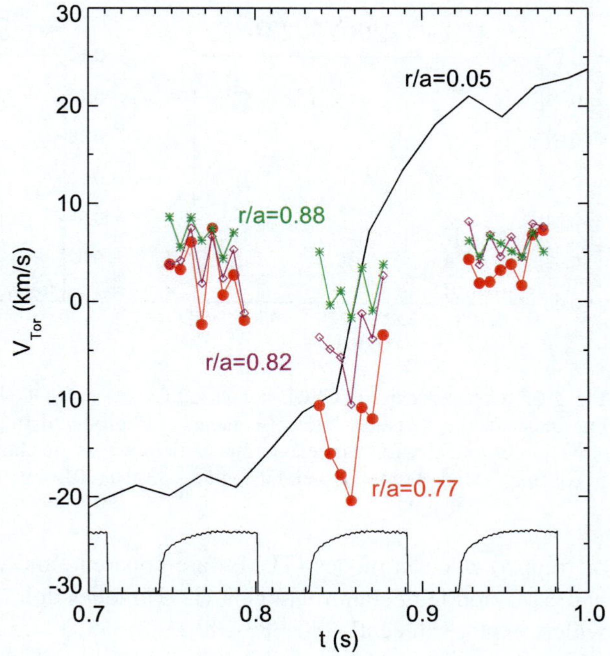

Fig. 5.57 The toroidal rotation velocity as a function of time at r/a = 0.05 (solid line), 0.77 (dots), 0.82 (diamonds) and 0.88 (asterisks), for a 3.0 T and 0.64 MA discharge. At the bottom are shown the diagnostic neutral beam pulses. Reproduced from Fig. 19 of [383] J.E. Rice et al. 'Observations of core toroidal rotation reversals in Alcator C-Mod ohmic L-mode plasmas', Nucl. Fusion **51**, 083005 (2011), with the permission of the IAEA

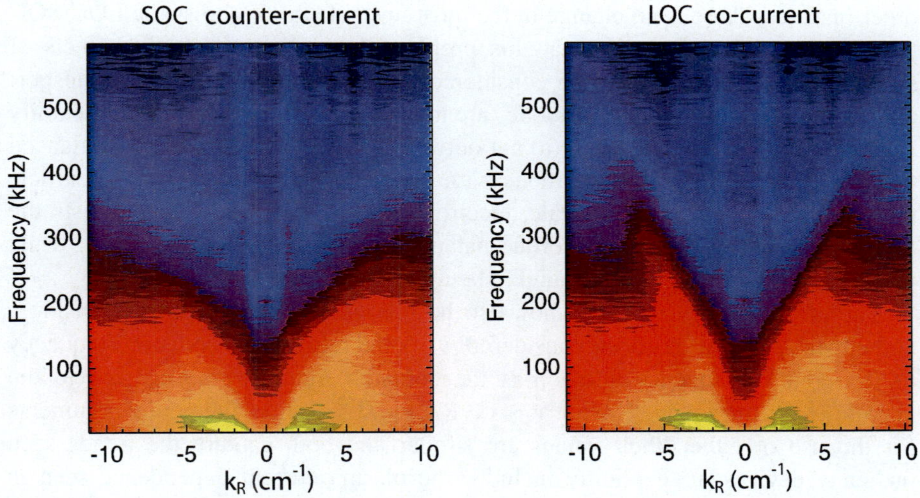

Fig. 5.58 Dispersion plots S(k,f) of turbulence before (0.608 s, counter-current rotation, left panel) and after (0.859 s, co-current rotation, right panel) the first rotation reversal in the C-Mod discharge shown in Fig. 5.38. Reprinted with permission from Fig. 3 of [384] J.E. Rice et al. 'Rotation Reversal Bifurcation and Energy Confinement Saturation in Tokamak Ohmic L-Mode Plasmas', Phys. Rev. Lett. **107**, 265001 (2011). Copyright (2011) by the American Physical Society

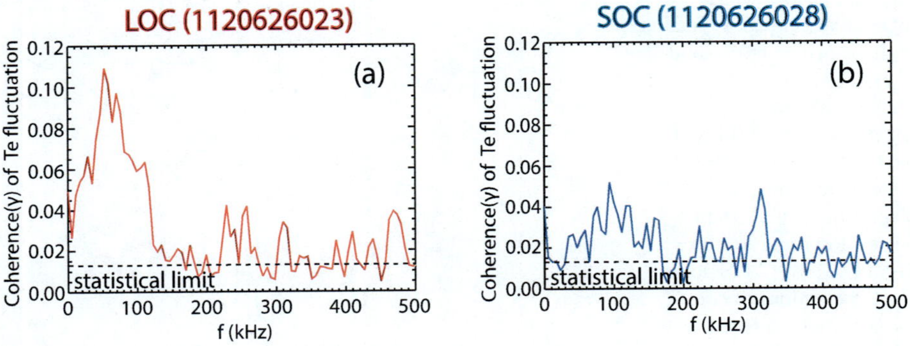

Fig. 5.59 Coherence of two CECE signals in LOC (**a**) and SOC (**b**) plasmas. The horizontal dotted line indicates the statistical limit of coherence. Reproduced from Fig. 3 of [412] C. Sung et al. 'Changes in core electron temperature fluctuations across the ohmic energy confinement transition in Alcator C-Mod plasmas', Nucl. Fusion **53**, 083010 (2013), with the permission of the IAEA

of trapped electron modes (TEMs) are apparent, and that with high collisionality and SOC, ion temperature gradient (ITG) modes dominate. This concept has been widely explored recently [80, 383–386, 388, 389, 394, 399–402, 411, 412, 414–417], often with accompanying gyro-kinetic simulations. In some cases these simulations show TEMs to be important at low ν_* and ITG modes dominant at high ν_*, while in other instances there is no change in the most unstable mode across the LOC/SOC transition. A very recent look at this problem demonstrates that the effects of sub-dominant modes need to be considered [417]. Using a quasi-linear transport approximation, four families of modes are identified, as shown in Fig. 5.61. Family I is ion-scale, ion-directed, with (a) net outward particle transport and (b) balanced transport. Family II is a hybrid mode with $k_y\rho_s \geq 1$ and a strong inward particle pinch. Family III is electron-scale, electron directed which mostly exhausts the electron heat flux. With SOC, torque balance is dominated by ITG modes while in the LOC regime, TEMs are found to be important.

The collisionality has been shown to be relevant in the LOC/SOC transition, and either ν_* or $\nu_{\rm eff}$ has been considered. ν_* is the ratio of the collision frequency to the bounce frequency and $\nu_{\rm eff}$ is the collision frequency normalized to the curvature drift frequency, with $\nu_{\rm eff} \propto \mathrm{RZ_{eff}n}_e/\mathrm{T}_e^2$. While there are arguments for the use of either, their values are similar and both capture the trends seen on many devices, ν_* explicitly includes the plasma current dependence seen in Figs. 5.40, 5.41, 5.48, 5.49, 5.50 and 5.60, and ν_* is directly connected to the effects of trapped electrons. The use of ν_* does a better job of organizing the critical reversals observed at different plasma current on C-Mod [385]. An excellent review of rotation reversals on various devices and the theoretical framework for an understanding of various mechanisms can be found in Ref.[418].

Scaling studies from AUG [388, 401] suggest that the electron density gradient is an important factor in determining the velocity gradient. This is demonstrated in Fig. 5.62 which shows the mid-radius intrinsic velocity gradient as a function of the

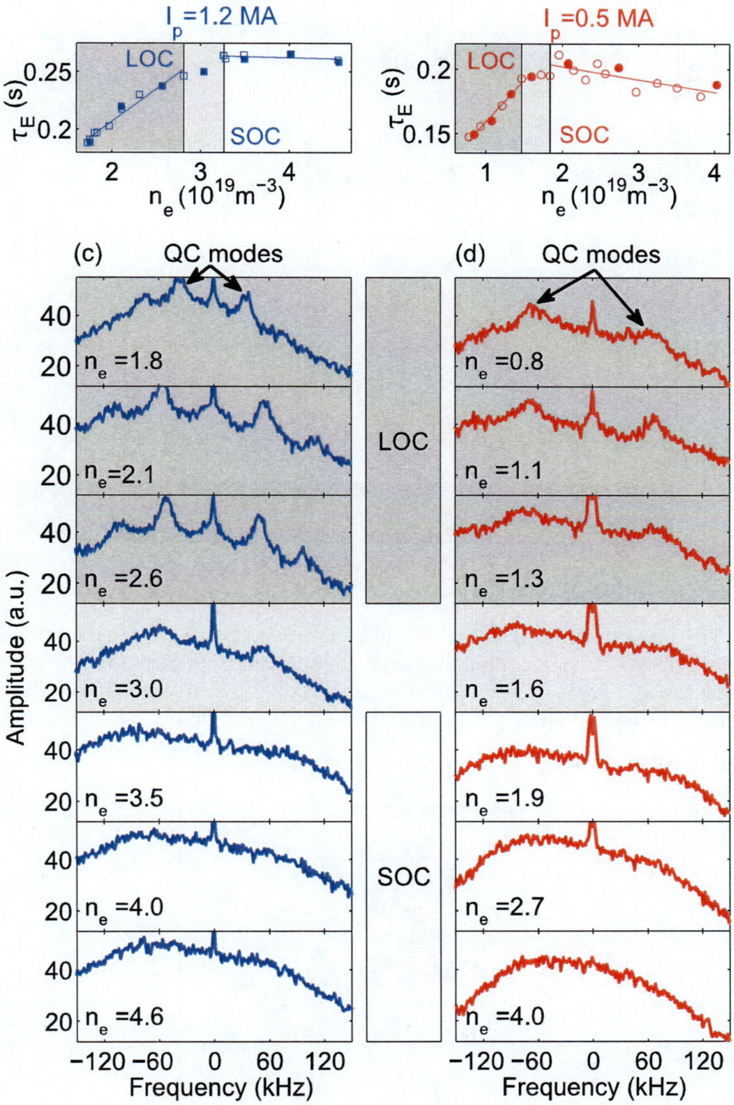

Fig. 5.60 Tore Supra reflectometry spectra during LOC/SOC transitions achieved at 3.8 T by scanning the central line averaged density n_e ($10^{19}/m^3$) at two different plasma currents: 1.2 MA (**a**, **c**) and 0.5 MA (**b**, **d**). At the top are the energy confinement times. Reproduced from Fig. 1 of [399] H. Arnichand et al. 'Quasi-coherent modes and electron-driven turbulence', Nucl. Fusion **54**, 123017 (2014), with the permission of the IAEA

local normalized density gradient in AUG L- and H-mode plasmas [401]. There is a strong correlation between the two. The steep gradient in the velocity is a good indicator of the sign of the core rotation, so these results indicate that the density gradient may be a proxy in the reversal process. While there is a gradual change

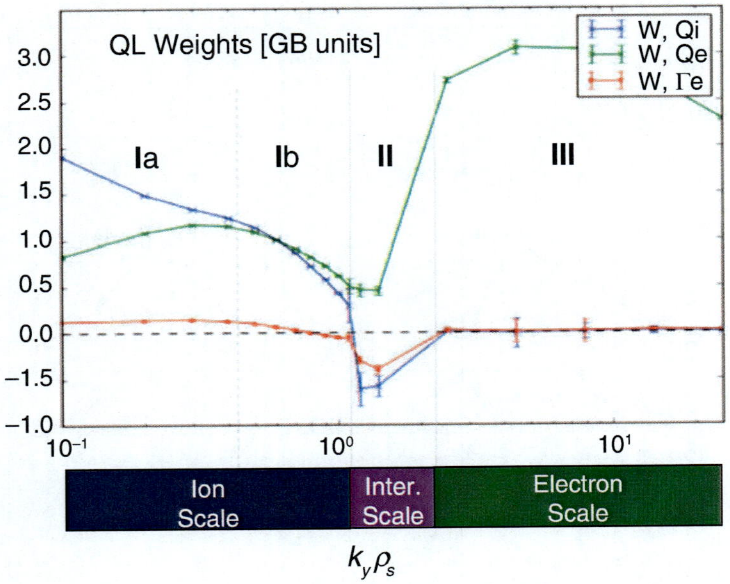

Fig. 5.61 The quasi-linear weights for the most unstable drift wave modes calculated from CGYRO for a 0.8 MA Ohmic C-Mod discharge. Reproduced from Fig. 4 from [417] N.M. Cao et al. 'Hysteresis as a Probe of Turbulent Bifurcation in Intrinsic Rotation Reversals on Alcator C-Mod', Nucl. Fusion **59** 104001 (2019), with the permission of the IAEA

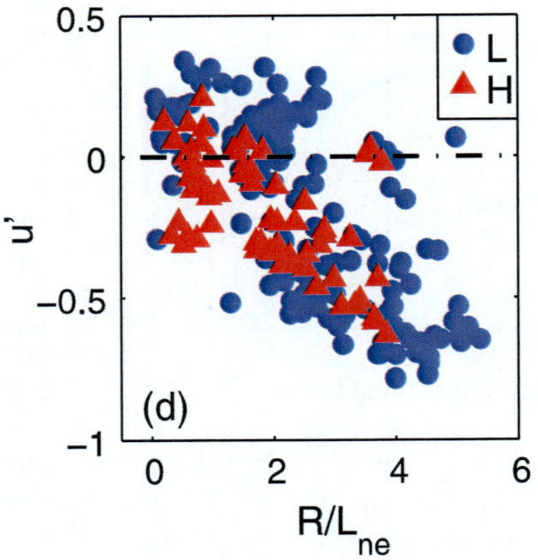

Fig. 5.62 The velocity gradient u′ as a function of the electron density gradient scale length, evaluated at r/a = 0.45, for L- (blue dots) and H-mode (red triangles) AUG plasmas. Reprinted with permission from Fig. 2 of [401] C. Angioni et al. 'Intrinsic Toroidal Rotation, Density Peaking, and Turbulence Regimes in the Core of Tokamak Plasmas', Phys. Rev. Lett. **107**, 215003 (2011). Copyright (2011) by the American Physical Society

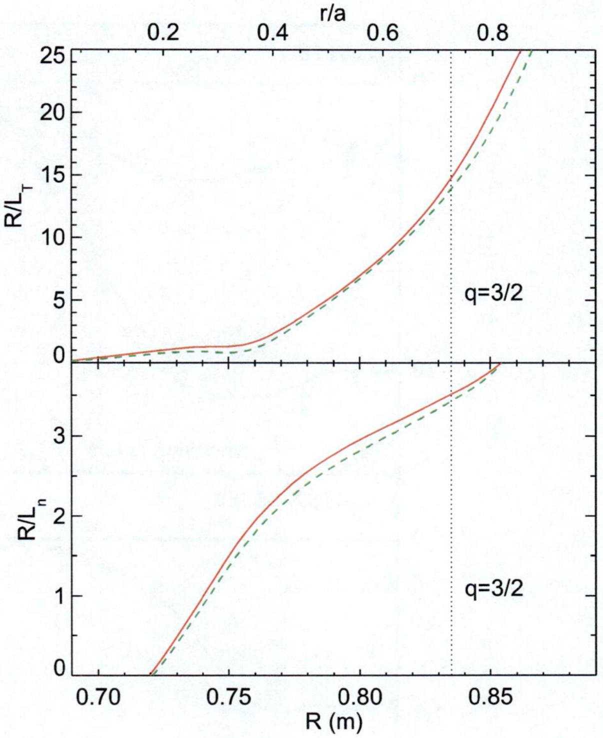

Fig. 5.63 The normalized electron temperature (top) and density (bottom) gradient profiles before (green dashed) and after (solid red) a rotation reversal in a 1.1 MA, 5.4 T C-Mod discharge. The vertical dotted line indicates the location of the q = 3/2 surface, just ouside of the reversal radius. Reproduced from Fig. 27 of [383] J.E. Rice et al. 'Observations of core toroidal rotation reversals in Alcator C-Mod ohmic L-mode plasmas', Nucl. Fusion **51**, 083005 (2011), with the permission of the IAEA

in the electron density scale length over a large range of collisionality peaking near the LOC/SOC transition in C-Mod plasmas (as was seen in the bottom frame of Fig. 5.50), there is no detectable change in the density gradient on either side of the rotation reversal [383, 419], or for that matter the temperature gradient. This can be seen in Fig. 5.63 which shows the temperature and density gradient scale length profiles before and after a rotation reversal in a C-Mod plasma [383]. There is no measurable difference in either the density or temperature gradients across the transition. This independence of the reversal process on background kinetic profiles calls into question the notion of a critical gradient being at play and suggests that something else is triggering the transition. Why there is a critical collisionality for this is intriguing.

Finally, it should be noted that some discharges exhibit the transformation of features such as the saturation of confinement and a reduction in turbulence across the LOC/SOC transition, including changes from flat to hollow velocity profile shapes, but without a change in sign of the rotation [391, 394, 395]. This is shown in the top frame of Fig. 5.64 for a KSTAR discharge with a density ramp to induce the LOC/SOC transition [391]. The profile shape changes look very similar to those seen in Fig. 5.54, flat in the LOC regime and hollow with SOC, except with everything shifted in the counter-current direction. The movement of the anchor point inward for lower plasma current is also similar to the results of

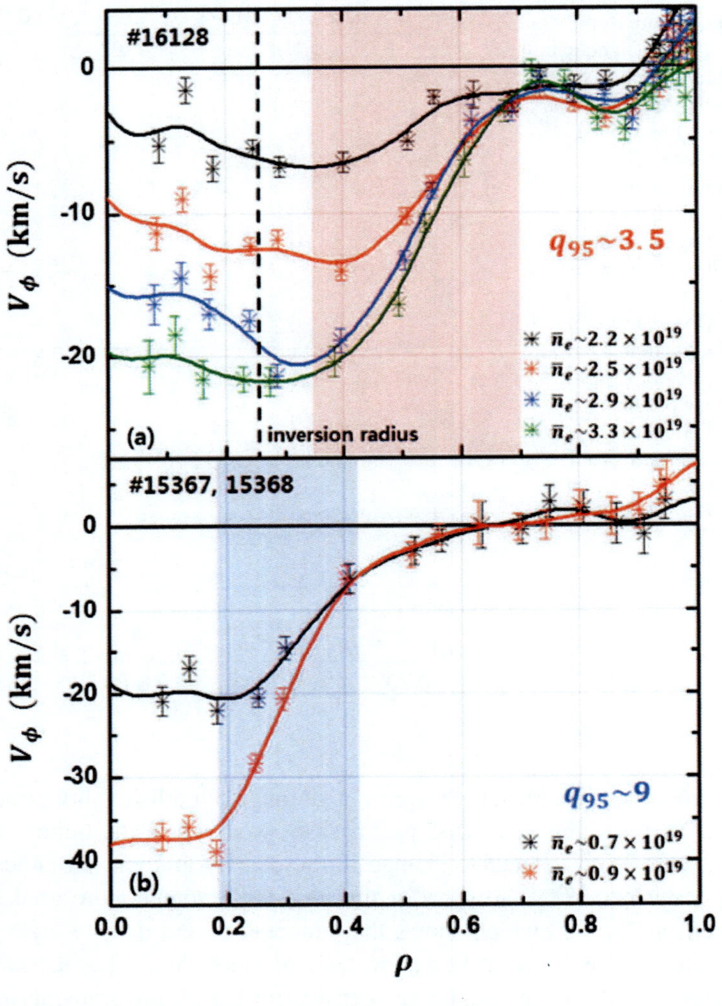

Fig. 5.64 Evolution of the intrinsic rotation profile during the rotation reversal (more general definition) in the KSTAR Ohmic plasmas: (**a**) w/ and (**b**) w/o sawtooth activity. Reproduced from Fig. 3 of [391] D.H. Na et al. 'A comprehensive study on rotation reversal in KSTAR: experimental observations and modelling', Nucl. Fusion **57**, 126008 (2017), with the permission of the IAEA

Fig. 5.55. It appears that there can be an offset velocity in Tore Supra and KSTAR discharges while all other features of the LOC/SOC transition are the same as seen on other devices. As has been documented in the earlier chapters, there are several mechanisms which can lead to an offset velocity, such as through ripple losses or NTV, for example. Application of RF heating power can also shift the velocity profiles and will be explored in the next section.

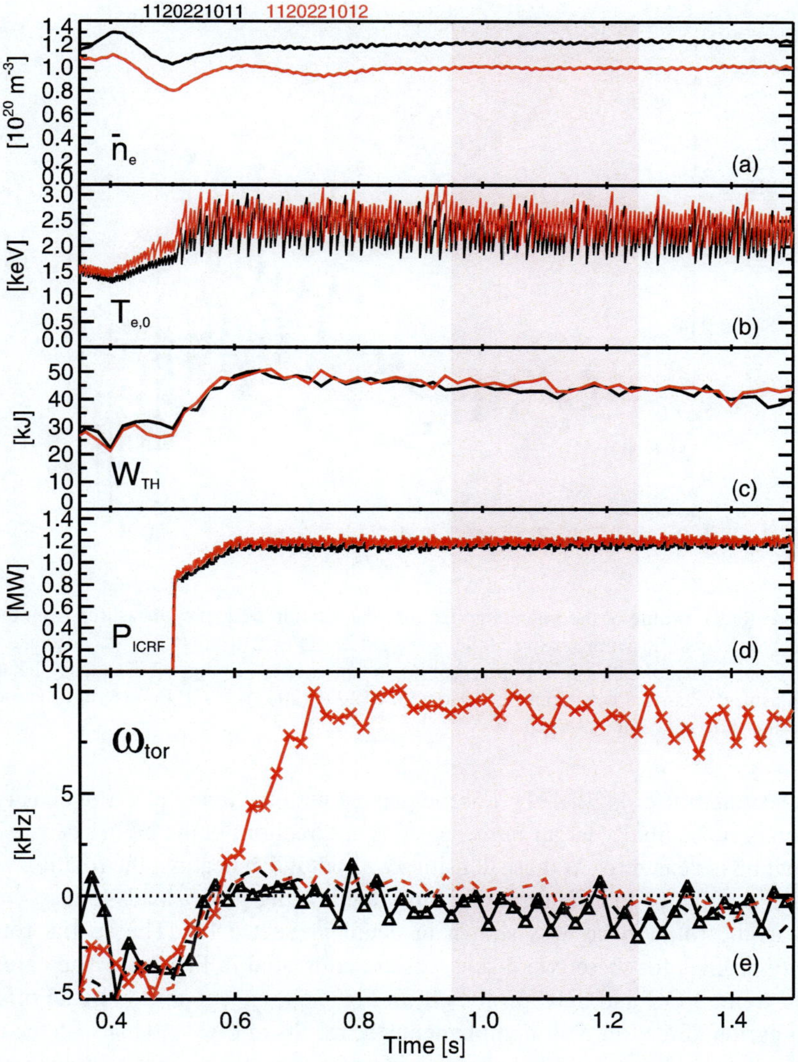

Fig. 5.65 Time history of two repeated 0.8 MA, 5.4 T plasmas with ∼20% difference in line averaged electron density (**a**). Both are heated with 1.2 MW of ICRF power (**d**) and the resulting stored energy (**c**) and core T_e (**b**) are very similar, but show dramatically different core rotation time histories (**e**). Fig. 1 from [420] M.L. Reinke et al. 'Density sensitivity of intrinsic rotation profiles in ion cyclotron range of frequency-heated L-mode plasmas', Plasma Phys. Control. Fusion **55**, 012001 (2013). ©IOP Publishing. Reproduced with permission. All rights reserved

5.2.4 Auxiliary Heated L-Mode

Rotation in L-mode plasmas with auxiliary heating can exhibit very similar behavior to that seen in purely Ohmic rotation reversals. Shown in Fig. 5.65 is a comparison of two C-Mod L-mode discharges, at slightly different density, with 1.2 MW of

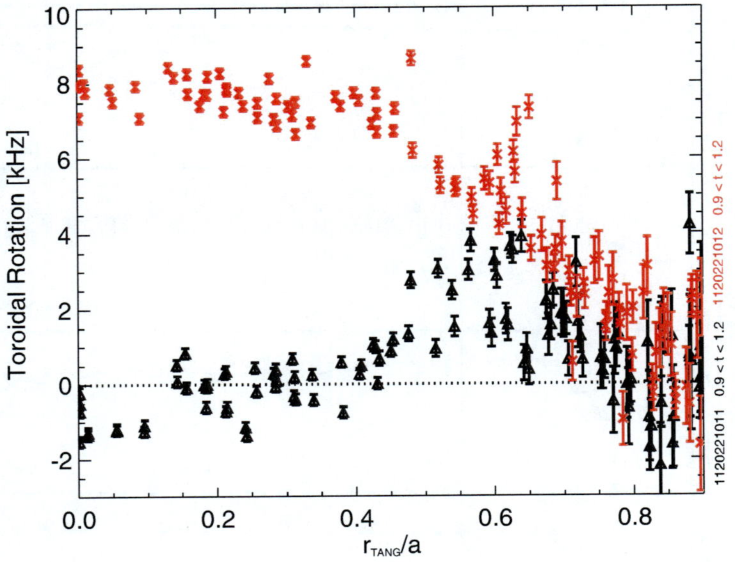

Fig. 5.66 Radial profile of the line integrated toroidal rotation frequency measured from the He-like Ar Doppler shift, averaged over $0.95 < t < 1.25$ s. Fig. 2 from [420] M.L. Reinke et al. 'Density sensitivity of intrinsic rotation profiles in ion cyclotron range of frequency-heated L-mode plasmas', Plasma Phys. Control. Fusion **55**, 012001 (2013). ©IOP Publishing. Reproduced with permission. All rights reserved

ICRF heating power [420, 421]. For the plasma with the lower electron density, the rotation is substantially larger in the co-current direction; in the higher density case the rotation is near zero. At these densities without ICRF heating, the rotation would have been in the counter-current direction; the heating power leads to a co-current offset to the rotation, as was shown in detail in Sect. 5.1.2. The spatial rotation velocity profiles for these two discharges are compared in Fig. 5.66. They are very similar to the LOC and SOC profiles from Fig. 5.54, except shifted by an offset in the co-current direction. This complements the results of Fig. 5.64 but with the offest in the opposite direction. These results indicate that the normal rotation reversal phenomenology still holds, but with an offset velocity. This is the motivation for adopting a more general definition of rotation 'reversal' which is governed by the mid-radius velocity gradient, even if there is not a sign change of the core velocity [391, 418].

This transition from peaked to hollow profile shape occurs at a higher density that would be expected for 5.4 T, 0.8 MA C-Mod discharges. However, since with the higher temperature due to the ICRF heating, the critical collisionality is still the same. If the phenomenology is the same as in Ohmic L-mode, the method of changing the collisionality through the temperature in auxiliary heated discharges allows expanded study of the reversal process. Collisionality changes of the rotation behavior with ECH have been investigated in AUG [115, 388] and KSTAR [340, 403], and in gyro-kinetic studies [422]. The effect of collisionless TEMs in flattening the rotation profile in ECH plasmas has also been explored [423].

5.2.5 Comments

For turbulent drive of intrinsic rotation, it is necessary to have some sort of symmetry breaking [6–8, 418]. A widely studied way of achieving this is through E×B shear which arises from pressure and/or temperature gradients prominent in enhanced confinement regimes, as was explored in Sect. 5.1. The GTS code has been used to simulate rotation reversals in DIII-D plasmas [392]. With SOC, there is a

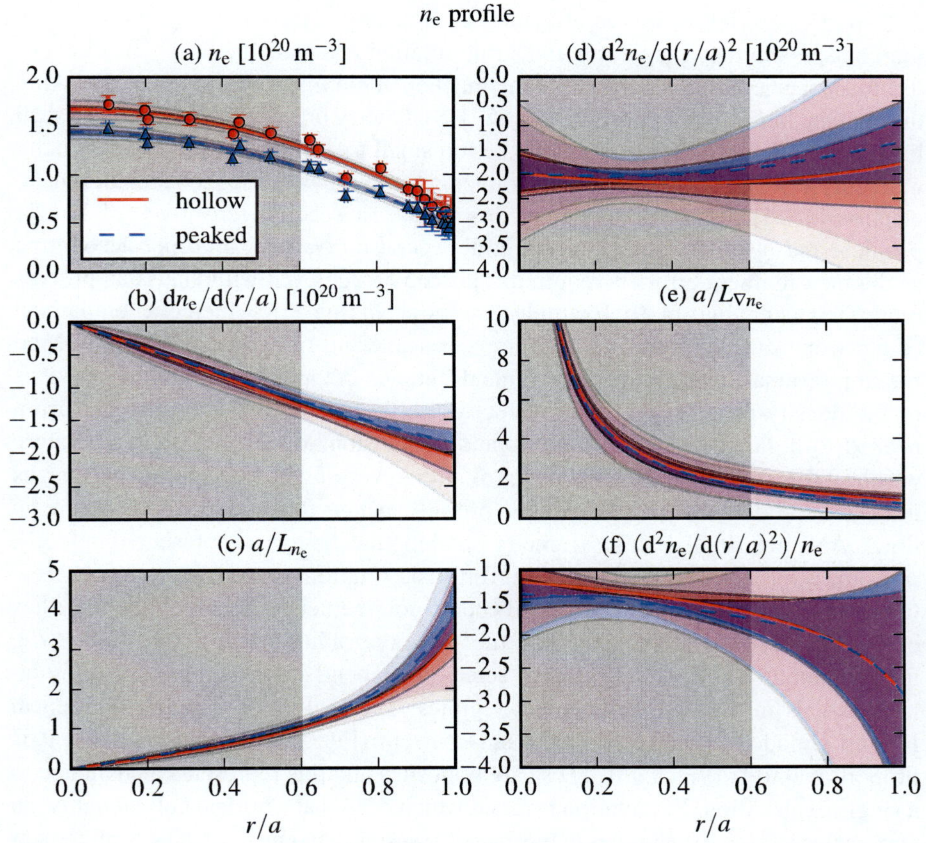

Fig. 5.67 Data and fitted n_e profiles for two C-Mod plasmas with different rotation profiles. Red circles and solid curves are from the higher density, hollow rotation profile plasma and blue triangles and dashed curves are from the lower density, peaked rotation profile discharge. For all curves, the dark band is $\pm 1\sigma$ and the light band is $\pm 3\sigma$. The shaded grey region r/a < 0.6 indicates the location of the largest difference in rotation. The subplots are: (**a**) the averaged data and fitted core profiles, (**b**) the first derivative of the profiles, (**c**) the normalized inverse gradient scale length of the profiles, (**d**) the second derivative of the profiles, (**e**) the inverse normalized gradient scale length of the gradient, and (**f**) the normalized second derivative. Reproduced from Fig. 3 of [419] M.A. Chilenski et al. 'Experimentally testing the dependence of momentum transport on second derivatives using Gaussian process regression', Nucl. Fusion **57**, 126013 (2017), with the permission of the IAEA

distinct positive peak of Π^{res} near mid radius which gives rise to the hollow rotation profile, and zonal flow shearing is the dominant symmetry breaking mechanism. With LOC, the residual stress profile is near zero, consistent with the observed flat rotation profile. Similar conclusions were found for AUG rotation reversals [424]. Another symmetry breaking mechanism is from eddy tilts due to electric and magnetic field shear, which can arise from changes in the magnetic configuration [376] (Sect. 5.2.1) or from external current drive [181–183] (5.2.2). In the former case, the shear induced Reynolds stress at the boundary drives the rotation and may be related to the H-mode power threshold difference with favorable and unfavorable drift [425]. A different source of symmetry breaking is through turbulent intensity gradients which may play a role in Ohmic rotation reversals [8, 426]. The effect of turbulent acceleration of particles along magnetic field lines as a symmetry breaking mechanism has also been considered [427]. Symmetry breaking due to neo-classical background flows [428] was found to be too small to account for observed velocity gradients in AUG [429]; larger flow gradients are found with the inclusion of kinetic electrons [424]. These diamagnetic flows rely on second derivatives of kinetic profiles, and in order to infer meaningful second derivatives, a proper accounting of the measurement errors is required. This can be achieved with Gaussian process regression techniques [419]. Examples are shown in Fig. 5.67 which is a comparison of electron density profiles from Thomson scattering on either side of a C-Mod rotation reversal. There is no discernible difference in the first or second derivatives of the density profiles, while the velocity profile has dramatically changed from peaked to hollow. A similar analysis and conclusions obtain for the temperature profiles. This rules out mechanisms for rotation reversal which are based on changes in the second deriviative of the kinetic profiles such as those based on diamagnetic flows. A summary of several symmetry breaking mechanisms is presented in [430].

Some approaches to understanding intrinsic rotation attempt to unify observations from both L-mode and enhanced energy confinement regimes. While there has been some success in characterizing the velocity gradient in this way [388, 401], the magnitude of the core velocity seems to depend upon completely different mechanisms in the two confinement regimes. The scaling of the core rotation in H- and I-mode is inverse to the plasma current (Fig. 5.7) while in L-mode it is proportional to I_p (Fig. 5.43). These are opposites and this represents a challenge to a single explanation. For rotation reversal which occurs at a critical collisionality, an approach unifying the changes in the energy transport channels, such as confinement saturation and the 'non-local' cutoff, would be prudent.

Chapter 6
Momentum Transport

As was shown in Eq. (2), the momentum flux consists of three terms, the momentum diffusivity, the momentum pinch and the residual stress. In this chapter will be presented how these transport coefficients are determined experimentally and how they are understood theoretically. The problem of obtaining these three is underdetermined since the observations consist of the time evolution of the velocity profile, which is not sufficient to fix three free parameters. In certain circumstances it is possible to isolate each of the three, or two of the three, and the results will be covered in Sects. 6.1, 6.2, and 6.3, respectively, with a comparison to theory in Sect. 6.4.

6.1 Momentum Diffusivity χ_ϕ

The most fundamental momentum transport coefficient is the diagonal term, the diffusivity, or viscosity, χ_ϕ. If the momentum source is well characterized, then it is possible to determine χ_ϕ by the response of the velocity profile through Eq. (1). The orignal and most common method is through the use of NBI [13, 14, 16, 54, 56, 60, 86–90, 92–95, 98, 103], while ignoring V_p and Π^{res}. The global momentum confinement time τ_ϕ can be determined either from the ratio of the steady state angular momentum density divided by the input torque, or from the rise (or decay) of the toroidal rotation following the turn on (turn off) of the neutral beam. The latter technique was first used on PLT [13] where τ_ϕ was found to be about 10 ms, two orders of magnitude lower than the neo-classical value, but very similar to the global energy confinement time. The first spatial profiles of χ_ϕ were determined from the measured velocity profiles in PDX [16] and are shown in Fig. 6.1. The core value of χ_ϕ is about 1 m^2/s, much larger than the neo-classical value of $\chi_\phi^{NC} = q^2 \nu_i \rho_i^2$ [431] which is around 0.01m^2/s. Similar experimental values for χ_ϕ, ranging from

J. Rice, *Driven Rotation, Self-Generated Flow, and Momentum Transport in Tokamak Plasmas*, Springer Series on Atomic, Optical, and Plasma Physics 119,
https://doi.org/10.1007/978-3-030-92266-5_6

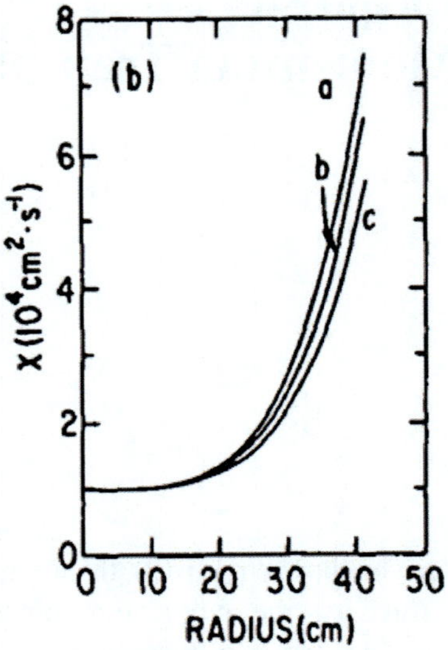

Fig. 6.1 Viscosity profiles which give the best fit to the data for different values of the charge exchange cross section. Reproduced from Fig. 12 of [16] K. Brau et al. 'Plasma rotation in the PDX tokamak', Nucl. Fusion **23**, 1643 (1983), with the permission of the IAEA

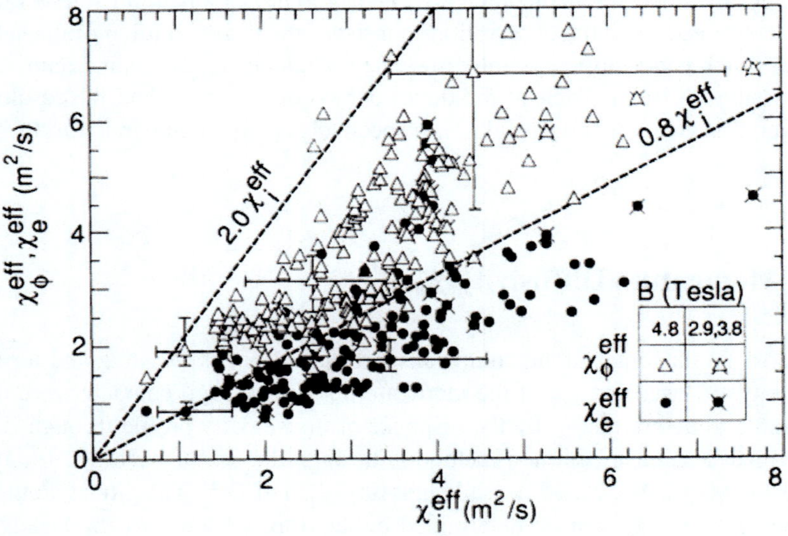

Fig. 6.2 Measured effective thermal and momentum diffusivities for all discharges in the experiment in the region $0.3 \leq r/a \leq 0.8$. Reprinted with permission from Fig. 3 of [86] S.D. Scott et al. 'Local measurements of correlated momentum and heat transport in the TFTR tokamak', Phys. Rev. Lett. **64**, 531 (1990). Copyright (1990) by the American Physical Society

0.1 to 2 m^2/s, have been found on other devices from NBI experiments, in different confinement regimes [14, 56, 86, 87, 89, 90, 94, 95, 103].

A parameter of interest is the Prandtl number $P_r = \chi_\phi/\chi_i$, the ratio of the momentum diffusivity to the ion thermal conductivity. A simple global determination of P_r can be found from τ_E/τ_ϕ, where τ_E is the energy confinement time. An example of the Prandtl number for TFTR discharges is shown in Fig. 6.2 [86]. P_r is seen to

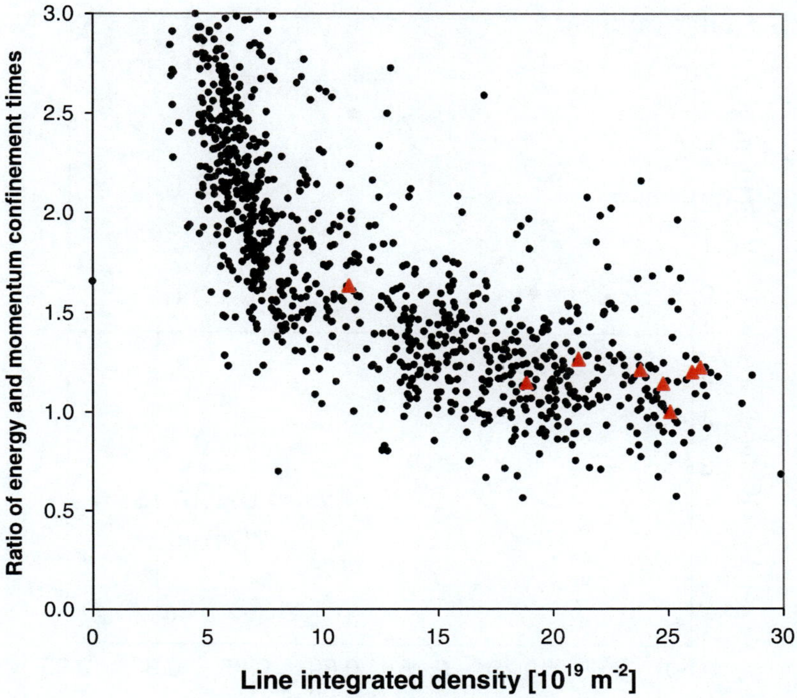

Fig. 6.3 The ratio of the energy confinement time to the momentum confinement time as a function of line integrated density. Determined for all types of discharges (optimized shear/ITB, L-mode, H-mode, hybrid) in the period 2000–2004, with NBI power above 6 MW, $I_p > 2$ MA. In the red triangles are shown high density ELMy H-mode discharges. Fig. 9 from [95] P.C. de Vries et al. 'Plasma rotation and momentum transport studies at JET', Plasma Phys. Control. Fusion **48**, 1693 (2006). ©IOP Publishing. Reproduced with permission. All rights reserved

vary between 0.8 and 2.0, as shown by the dashed lines. Prandtl numbers between 0.1 and 3 have been found in a variety of devices [13, 14, 16, 54, 56, 60, 86–90, 92–95, 98, 103] and this indicates the close relation between momentum and energy confinement. Scalings of P_r have been performed and an example is shown in Fig. 6.3 from JET [95]. This global estimate of P_r is seen to decrease at higher density.

A drawback of using NBI to determine χ_ϕ is that not all of the input torque is from the beams; significant intrinsic torque was found in NBI discharges as was seen in Figs. 5.18 and 5.19, so care must be taken when using beam injection to account for all torques. One way around this is to use the intrinsic torque generated at the L-H transition to find χ_ϕ. Taking advantage of the fact that the transition is instantaneous and originates in the pedestal region, the momentum diffusivity can be determined by the evolution of the core rotation [35, 36, 71]. Figure 6.4 shows the evolution of the toroidal rotation (top frame) at four radii following an L-H transition at 0.66 s for a C-Mod discharge [71]. In the bottom frame are the simulated results,

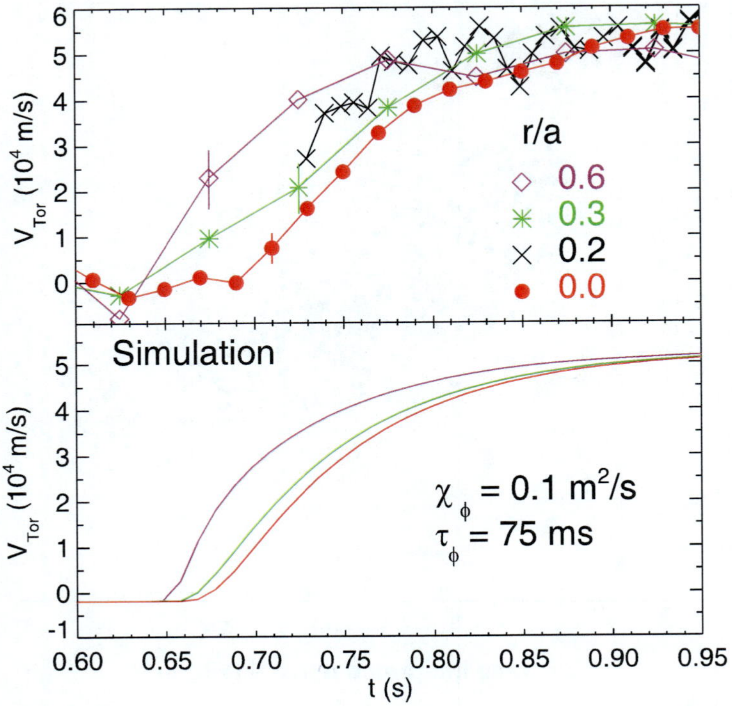

Fig. 6.4 The time evolution of the toroidal rotation velocity at four spatial locations for a C-Mod H-mode discharge (top frame). In the bottom frame are simulations with χ_ϕ = 0.1 m^2/s. Reproduced from Fig. 2 of [71] J.E. Rice et al. 'Toroidal rotation and momentum transport in Alcator C-Mod plasmas with no momentum input', Phys. Plasmas **11**, 2427 (2004), with the permission of AIP Publishing

assuming an instantaneous edge source, with $\chi_\phi = 0.1$ m^2/s, in good agreement with the observations.

Another method of velocity perturbation at the plasma edge, from varying the X-point location, was shown in Fig. 5.30. Again, χ_ϕ can be determined from the core velocity evolution following a jog in SSEP (which is an edge pertrubation), as shown in Fig. 6.5 for a C-Mod L-mode discharge. After the change from LSN to USN at 0.68 s, the velocity response was first seen at the edge, then propagating into the center. Modeling for this evolution indicates $\chi_\phi \sim 1$ m^2/s, larger than that in H-mode. A larger momentum diffusivity in L-mode compared to H-mode was first noted in ASDEX plasmas [87], and is consistent with the increase in energy confinement.

The previous discussion assumed that the momentum transport is purely diffusive. It has been long known that there is evidence for non-diffusive behavior [60, 89, 90] and since the momentum pinch is an obvious candidate in an analogy to Fick's law, it will be discussed in the next section.

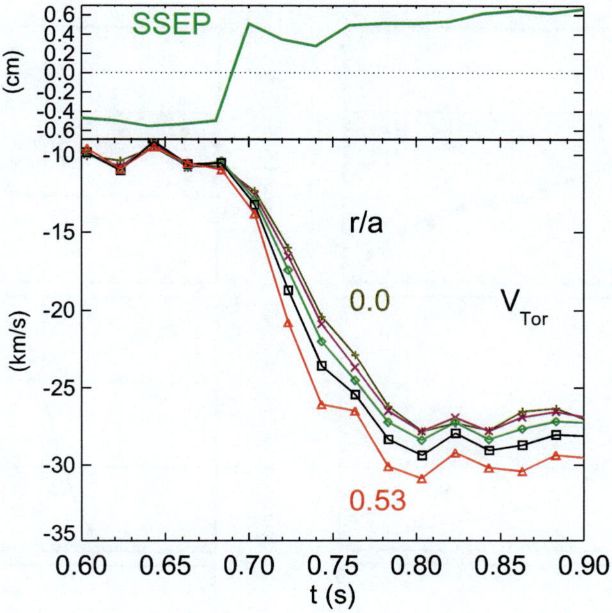

Fig. 6.5 The evolution of the toroidal rotation at different radii, evenly spaced between r/a of 0.0 and 0.53 (bottom frame) following at switch from LSN to USN at 0.68 s (top frame)

6.2 Momentum Pinch V_p

The first direct evidence for a momentum pinch came from a comparison of velocity profiles in JT-60U plasmas with on- and off-axis NBI [89]. Shown in Fig. 6.6 are the measured toroidal rotation profiles (top), calculated NBI momentum input (middle) and inferred momentum diffusivity profiles (bottom) for discharges with on-axis (left) and off-axis (right) beams. The flat central rotation profile with off-axis momentum input (right) infers an extremely small central diffusivity, and the only way the off-axis momentum input can be transported inward is through an inward momentum pinch. $V_p \sim 1$–3 m/s is required to account for the velocity profile, which is confirmed by Fourier analysis of beam modulation experiments. Further evidence of an inward pinch was found in C-Mod ELM-free H-modes following the L-H transition [35, 36] (see Fig. 1.15). The profile peaking demonstrated in Fig. 6.7 from an edge source cannot be due solely to momentum diffusivity. Modeling with $V_p = -12$ (r/a) m/s and $\chi_\phi = 0.4$ m^2/s, shown by the curves, is in good agreement with observations.

Fourier analysis of beam modulation experiments, as was seen in Fig. 2.2, can also be used to determine V_p and χ_ϕ profiles [89, 97, 99, 102, 106, 110–112, 114, 116]. An example for JT-60U plasmas is shown in Fig. 6.8 [97]. This inward pinch and momentum diffusivity are similar to those seen in other devices, in addition to

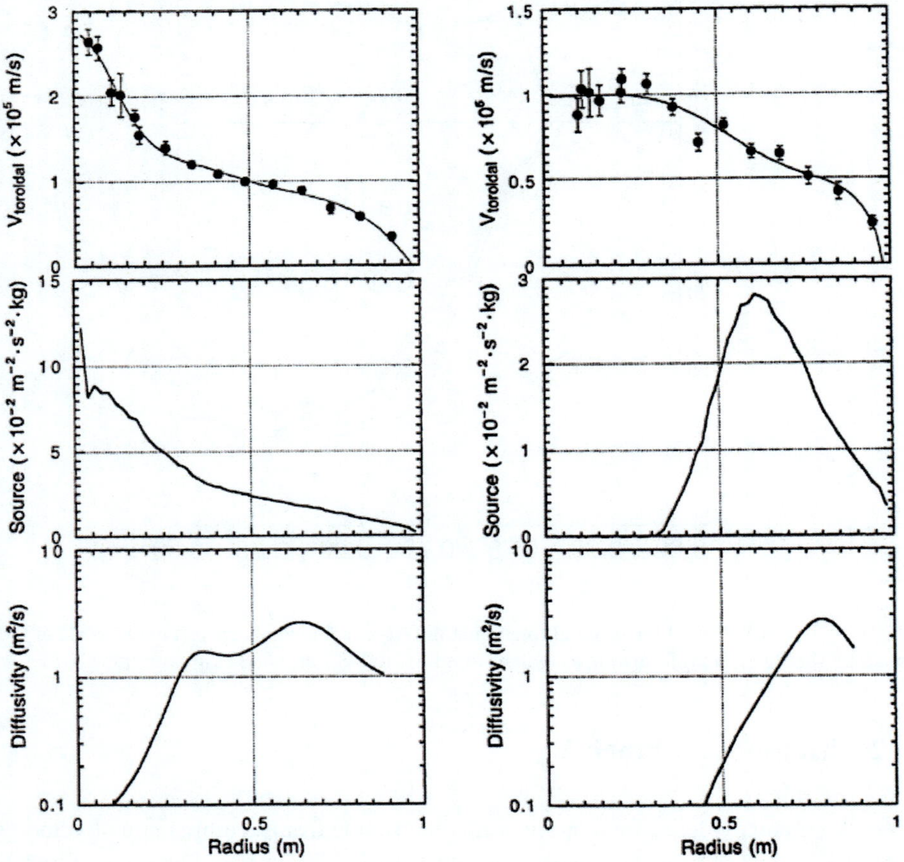

Fig. 6.6 Radial profiles of measured toroidal velocity (top), momentum source (middle) and momentum diffusivity (bottom) for on-axis (left) and off-axis (right) beam injection. Reproduced from Figs. 2 and 3 of [89] K. Nagashima et al. 'Experimental determination of non-diffusive toroidal momentum flux in JT-60U', Nucl. Fusion **34**, 449. (1994), with the permission of the IAEA

those inferred from modulating perturbation coils [104, 109, 432] or the toroidal magnetic field ripple [191].

A convenient dimensionless parameter for characterizing the momentum convection is the pinch number, RV_p/χ_ϕ, where R is the major radius. This parameter has been studied on many devices [111, 116–118, 335, 432] and typical values are between -1 and -10. An example of a scaling study is presented in Fig. 6.9 which shows the dependence of the pinch number on the normalized density gradient, R/L_n for JET discharges [116]. The sign and magnitude of the observed pinch is captured by the Coriolis pinch model [433] which yields $-RV_p/\chi_\phi = 4 + R/L_n$. There is also fair agreement with gyro-kinetic simulations, which has been observed in other studies [111, 117, 118, 432].

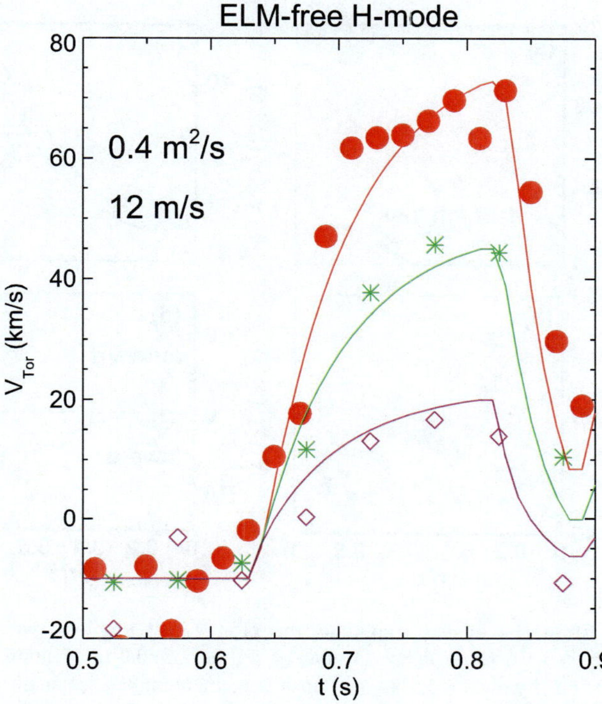

Fig. 6.7 Impurity toroidal rotation velocity at three radii (red dots, green asterisks and purple diamonds for r/a = 0.0, 0.3 and 0.6, respectively) for an ICRF heated ELM-free H-mode discharge. Simulations with χ_ϕ = 0.4 m^2/s and V_p = 12 m/s are shown as the color-coded curves for each radius. Reprinted with permission from Fig. 4 of [35] W.D. Lee et al. 'Observation of Anomalous Momentum Transport in Tokamak Plasmas with No Momentum Input', Phys. Rev. Lett. **91**, 205003 (2003). Copyright (2003) by the American Physical Society

In an effort to document the behavior of the momentum pinch and diffusivity, with an eye toward a fundamental understanding of the underlying processes, a multi-machine database has been constructed [120], including the devices C-Mod, DIII-D, JET, NSTX and JT-60U. An example of the results of the scaling study is presented in Fig. 6.10. A Prandtl number close to unity over a large range is evident in (a) and the magnitude of the inward pinch is well correlated with χ_ϕ in (b).

There is evidence that addition of ECH can alter momentum transport coefficients [332, 335]. This effect is demonstrated in Fig. 6.11 which shows χ_ϕ and V_p profiles for JT-60U H-mode discharges with and without ECH [332]. With ECH there is an increase in magnitude of both χ_ϕ and the inward pinch. However, the effect on momentum transport due to a change in Π^{res} from ECH might be another explanation for these differences.

The deviation of the observed pinch from that expected from the Coriolis pinch model and the dominant behavior of the residual stress (Chap. 5) in some circumstances, coupled with the fact that the momentum transport problem is

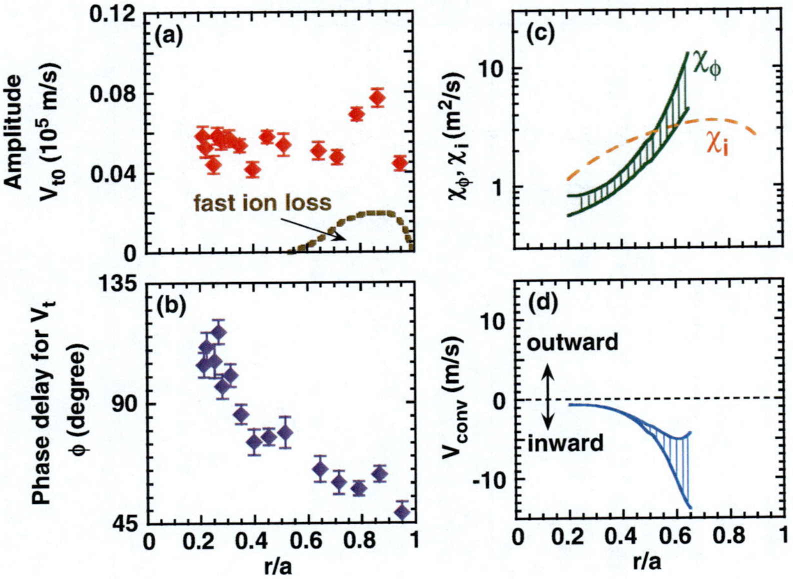

Fig. 6.8 Profiles of (**a**) modulated amplitude, the prompt fast ion loss, which works as a momentum source, and (**b**) phase delay. Profiles of (**c**) the toroidal momentum diffusivity and (**d**) the convection velocity obtained from the transient momentum balance analysis. The ion heat diffusivity from the power balance is also shown. Reproduced from Fig. 4 of [97] M. Yoshida et al. 'Momentum transport and plasma rotation profile in toroidal direction in JT-60U L-mode plasmas', Nucl. Fusion **47**, 856 (2007), with the permission of the IAEA

underdetermined, suggest that the effects of Π^{res} on momentum transport should be examined- the subject of the next section.

6.3 Residual Stress Π^{res}

Early evidence for off-diagonal components of the momentum flux from beam modulation studies showed that this term was not proportional to the value of the velocity [90, 91], thereby not consistent with a momentum pinch. Furthermore, analysis of observed velocity profiles on JT-60U could not be accounted for using only the χ_ϕ and V_p profiles determined from beam modulation experiments [101, 119]. An additional term, proportional to the pressure gradient, was required to match the measured v_ϕ profiles. An example is demonstrated in Fig. 6.12 which shows observed velocity profiles for JT-60U H-mode plasmas, in addition to the calculated profiles based on the two transport coefficients χ_ϕ and V_p determined from modulation experiments. A much better fit to the data requires inclusion of a Π^{res} term proportional to the pressure gradient. This is an unequivocal demonstration of the role of the residual stress in momentum transport.

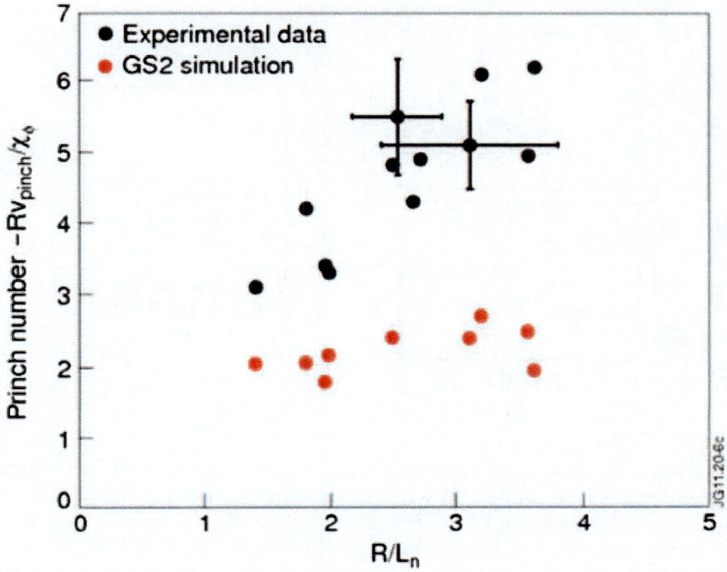

Fig. 6.9 Experimental (black dots) and simulated (red dots) pinch numbers, $-\mathrm{RV}_p/\chi_\phi$, averaged over the range $0.4 < \rho < 0.8$, as a function of the inverse density gradient length R/L$_n$. Reproduced from Fig. 6 of [116] T. Tala et al. 'Parametric dependences of momentum pinch and Prandtl number in JET', Nucl. Fusion **51**, 123002 (2011), with the permission of the IAEA

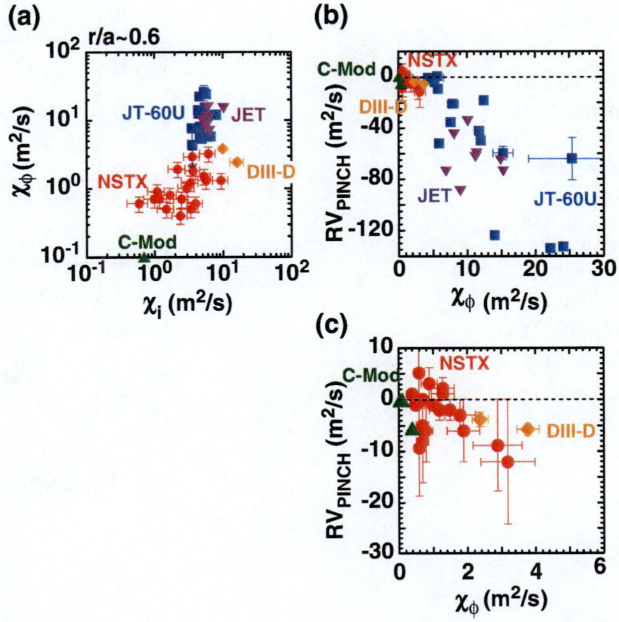

Fig. 6.10 (**a**) Correlation between χ_ϕ and the heat diffusivity (χ_i) in five machines. (**b**) Correlation between χ_ϕ and V_p at r/a = 0.6. (**c**) An enlarged view of (b) for C-Mod, DIII-D and NSTX. Reproduced from Fig. 3 of [120] M. Yoshida et al. 'Momentum transport studies from multi-machine comparisons', Nucl. Fusion **52**, 123005 (2012), with the permission of the IAEA

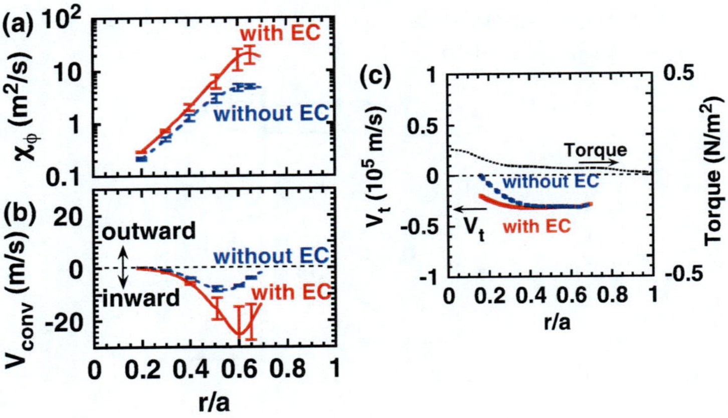

Fig. 6.11 Profiles of (**a**) the toroidal momentum diffusivity (χ_ϕ) and (**b**) the pinch velocity (V_{conv}) with ECRH (solid lines) and without ECRH (dashed lines) in H-mode plasmas. (**c**) Profiles of the calculated v_ϕ with and without ECRH from the momentum equation with χ_ϕ and V_{conv}. The torque input profile is also shown. Reprinted with permission from Fig. 2 of [332] M. Yoshida et al. 'Rotation Drive and Momentum Transport with Electron Cyclotron Heating in Tokamak Plasmas', Phys. Rev. Lett. **103**, 065003 (2009). Copyright (2009) by the American Physical Society

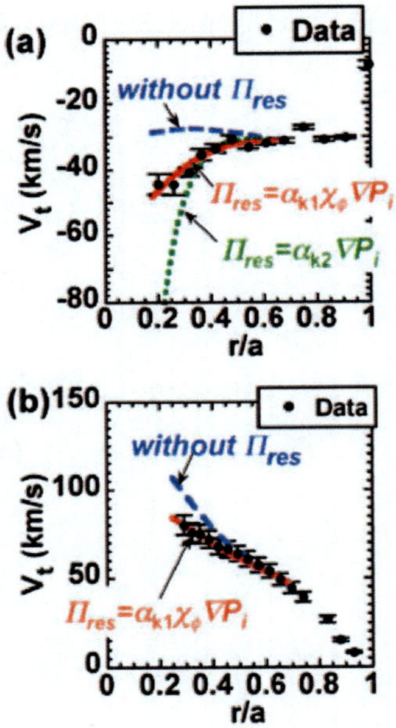

Fig. 6.12 Profiles of measured toroidal velocity (dots) compared to calculated profiles without Π^{res} (dashed lines) and with Π^{res} proportional to the pressure gradient (solid orange lines) for H-modes with (**a**) balanced beams and (**b**) co-injected beams. Reproduced from Fig. 4 of [119] M. Yoshida et al. 'Core and edge toroidal rotation study in JT-60U', Nucl. Fusion **52**, 023024 (2012), with the permission of the IAEA

6.4 Comments

Experimental values for the momentum diffusivity in the plasma core range from 0.1–2 m^2/s. This is much, much larger than the classical diffusivity due to collisions, $\nu_i \rho_i^2$, indicating that χ_ϕ is dominated by turbulent processes. Experiments have also shown that the momentum diffusivity is tied to the ion thermal conductivity, with Prandtl numbers χ_ϕ/χ_i varying from 0.1–2. An early fluid treatment of ITG mode turbulence demonstrated that χ_ϕ and χ_i are identical [434]. If resonant scattering by slightly suprathermal ions is taken into account, P_r is in the range from 0.2 to 0.5 [11]. Detailed parameter scans in linear and non-linear gyro-kinetic studies have confirmed that P_r is of order unity [6, 435, 436].

Momentum convection velocities have been found in experiments to be in the range from -2 to -30 m/s, directed inwards. There have been many theoretical studies of the momentum pinch [6, 7, 11, 357, 433, 437–443]. In general V_p can be decomposed into a turbulent equipartition (TEP) piece and a thermo-electric (TH) part. The TEP component is purely inward and its origin is in the compressibility of the $\mathbf{E}\times\mathbf{B}$ flow in toroidal geometry [11]. As was mentioned in Sect. 6.2, the dimensionless pinch number, $-RV_p/\chi_\phi$, is a convenient form for comparison with theory. The Coriolis pinch [6, 7, 433] arises from broken symmetry due to toroidal flow when treated in a co-rotating frame and gives $-RV_p/\chi_\phi = 4 + R/L_n$. This captures qualitatively the experimental results shown in Fig. 6.9 for JET plasmas in a scaling with R/L_n. Comparison with the experimental $-RV_p/\chi_\phi$ radial profile for a C-Mod H-mode discharge (with no momentum input) is shown in Fig. 6.13.

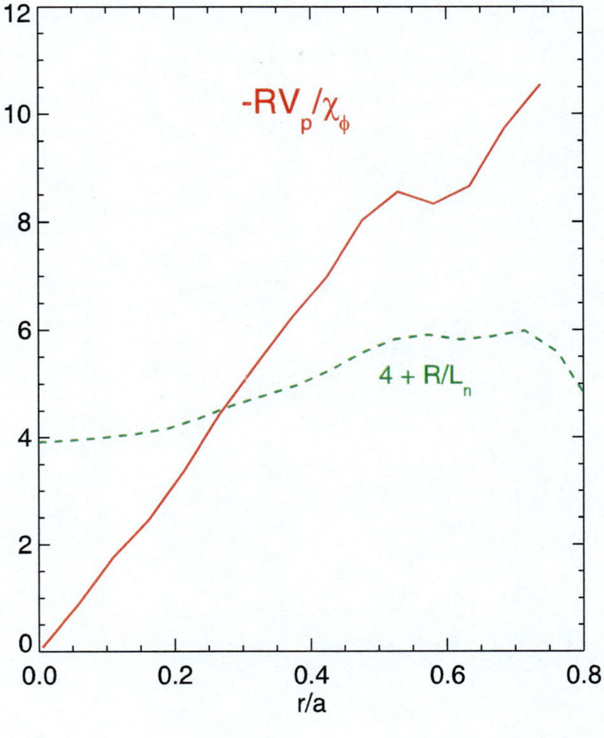

Fig. 6.13 The radial profile of the pinch parameter (solid red) for a C-Mod H-mode discharge compared to the Coriolis pinch model (green dashed)

The calculated profile is everywhere within a factor of two, and suggests that there might be another mechanism at play, possibly the residual stress. Comparison of the three terms of the momentum flux in Eq. (2) for intrinsic rotation in C-Mod discharges reveals that they are all of similar magnitude. For the first term, using χ_ϕ = 1 m^2/s and a typical velocity gradient in H-mode of 5×10^4 m/s over 0.1 m gives 5×10^5 m^2/s^2. For the second term, a momentum pinch of 10 m/s with a typical velocity of 5×10^4m/s yields 5×10^5 m^2/s^2. For the third term with a typical intrinsic torque density of 1 N/m^2 (equal to $\nabla \cdot \Pi^{res}$), with R = 0.67 m, a density of $10^{20}/m^3$ and a gradient scale length of 0.1 m for Π^{res} also gives 5×10^5 m^2/s^2. The main point here is that all three terms are potentially equally important when considering momentum transport, and that ignoring Π^{res} can lead to large uncertainties in the momentum flux. Recent studies in AUG plasmas have succeeded in isolating Π^{res} from a variety of modulation experiments [444].

Chapter 7
Discussion and Future Outlook

Based upon the results presented in Chaps. 2–5, it is fair to say that the understanding of observed rotation in tokamak plasmas is quantitative at best and not even qualitative at worst. Even in the most fundamental example of direct external drive (Sect. 2.1.1), a substantial fraction of the rotation observed following NBI is not in fact due to direct drive from the beams. As has been shown in DIII-D H-mode plasmas [100], the intrinsic torque is as large as that from one neutral beam source. This calls into question the traditional method of determining momentum transport coefficients from observed rotation profiles assuming momentum input (calculated) from the beams. In the case of ICRF waves (Sect. 2.1.2) in the minority heating scheme, observations show rotation in both directions, with complicated profile shapes [142, 156–158, 162], and agreement with theoretical models [142] is not even qualitative. Regarding ICRF MCFD (Sect. 2.1.2.2), some observations [136–138] agree with theory in direction, and in scaling with ^{3}He fraction, magnetic field and ICRF frequency, while in other instances [141] the direction is wrong. These results indicate that momentum input from RF waves is not well understood at all and that indirect effects from intrinsic torque due to the residual stress enabled by RF waves is dominant. Regarding direct momentum input from LH waves (Sect. 2.1.3), there is some quantitative agreement between observed rotation (direction, magnitude and scaling with power) [172] only if there is not substantial driven current. Once the q profile is modified, the observed rotation is in the opposite direction to the momentum input from the LH waves. In some cases there are observed effects in the core rotation when the LH waves cannot even penetrate past the LCFS [169], indicating the complication of understanding wave induced momentum input. For momentum sinks due to locked modes, magnetic braking and NTV (Sect. 3.2), the agreement between experiment and theory is often very good [253].

Numerous examples of the comparison between observations and the predictions of neo-classical theory were shown in Chap. 4. A huge range of agreement/disagreement was demonstrated, from excellent quantitative comparisons (e.g. Fig. 4.2) to complete disparity (e.g. Fig. 4.6). It must be that in the cases of

J. Rice, *Driven Rotation, Self-Generated Flow, and Momentum Transport in Tokamak Plasmas*, Springer Series on Atomic, Optical, and Plasma Physics 119,
https://doi.org/10.1007/978-3-030-92266-5_7

disagreement, other effects not taken into account are dominant. This indicates the importance of the residual stress, summarized in Chap. 5. It is fair to say that the understanding of intrinsic rotation is qualitative at best and that quantitative calculations of rotation due to the residual stress is sorely lacking. The problem is extremely complicated and comprehensive modeling, taking into account the many phenomena summarized in Chap. 5, needs to be undertaken. On top of the mixed results with respect to disparity between experiment and theory regarding sources, there is only qualitative agreement of the transport coefficients, as discussed in Chap. 6. It is essential to include Π^{res} when characterizing momentum transport.

In order to demonstrate active control of the velocity profile in tokamak plasmas, understanding of sources (direct and indirect) and sinks, including momentum transport, is necessary. In practice, all known effects must be taken into account including rotation behavior within the neo-classical framework. As was seen in Chap. 4, there are numerous instances where other effects dominate neo-classical considerations. Given that MHD equilibria can influence rotation (Sect. 5.2.1) and that turbulence and rotation are sometimes observed to change together (Sect. 5.2.3), a proper approach is to ensure rigorous integration of neo-classical and MHD concepts with the influence of turbulence, as outlined in Ref. [445], with attempted implementation for example in Ref. [166]. Furthermore, the turbulence itself depends upon whether the dominant modes are driven by electrons or ions, and interaction between these modes is important to consider [446].

There are many tools available for velocity profile tailoring. Direct rotation drive from NBI has been documented and counter-current rotation from LH waves has been shown. For ICRF minority heating and mode conversion flow drive, rotation in both directions has been seen. Regardless of the direct source, it is important to account for particle orbit and neo-classical effects. Rotation braking from magnetic perturbations has also been observed. Indirect sources from the residual stress have been demonstrated, either through temperature and density gradients, magnetic configuration, collisionality or the q profile. Changes in rotation through the collisionality have been achieved by varying the density, plasma current, magnetic field, electron temperature, impurity content, as well as the major radius and aspect ratio. By utilizing these various effects, where relevant, the prospect for controlling the rotation profile is within reach.

Areas for future work include explanation of MCFD observed in both directions, the role of kinetic stresses and quantitative modeling of the residual stress and its influence on rotation profiles. A key to understanding Π^{res} lies in the direct measurement of velocity fluctuations. With respect to rotation reversals, why they occur abruptly at a critical value of nqR/B is not understood. There is a tantalizing suggestion that this may be related to the density limit, with a value of $n/n_G \sim 1/2$. The connexion among rotation reversal, the LOC/SOC transition, the 'non-local' cut-off and disappearance of high k turbulent features may provide important clues. Many explanations have been put forward for each of these phenomena separately with mixed success; a good path forward may be an integrated approach addressing all simultaneously. There are many open questions which need to be resolved before a claim of comprehensive understanding of rotation reversals can be made. Why is the reversal transition so sharp at the critical nqR/B? There is demonstrated an hysteresis of 10% in the density following up and down ramps, which evolves on a momentum confinement time scale. What determines the reversal stagnation point? Is there a connexion between the rotation reversal critical nqR/B and the H-mode power threshold [447]?

References

1. Ida K., 1998 *Plasma Phys. Control. Fusion* **40** 1429.
2. Rice J.E., 2008 *J. Phys. Conf. Ser.* **123** 012003.
3. deGrassie J.S., 2009 *Plasma Phys. Control. Fusion* **51** 124047.
4. Ida K. and Rice J.E., 2014 *Nucl. Fusion* **54** 045001.
5. Rice J.E., 2016 *Plasma Phys. Control. Fusion* **58** 083001.
6. Peeters A.G. *et al.*, 2011 *Nucl. Fusion* **51** 094027.
7. Angioni C. *et al.*, 2012 *Nucl. Fusion* **52** 114003.
8. Diamond P.H. *et al.*, 2013 *Nucl. Fusion* **53** 104019.
9. Coppi B. and Zhou T., 2014 *Nucl. Fusion* **54** 093001.
10. Parra Felix I. and Barnes Michael, 2015 *Plasma Phys. Control. Fusion* **57** 045002.
11. Diamond P.H. *et al.*, 2009 *Nucl. Fusion* **49** 045002.
12. Bell M.G., 1979 *Nucl. Fusion* **19** 33.
13. Suckewer S. *et al.*, 1979 *Phys. Rev. Lett.* **43** 207.
14. Suckewer S. *et al.*, 1981 *Nucl. Fusion* **21** 1301.
15. Kostek C.A. *et al.*, 1983 *Plasma Phys.* **25** 421.
16. Brau K. *et al.*, 1983 *Nucl. Fusion* **23** 1643.
17. Bugarya V.I. *et al.*, 1985 *Nucl. Fusion* **25** 1707.
18. Mattioli M. *et al.*, 1988 *J. Appl. Phys.* **64** 3345.
19. Duval B.P. *et al.*, 1992 *Nucl. Fusion* **32** 1405.
20. Rowan William L. *et al.*, 1992 *Phys. Fluids B* **4** 917.
21. Severo J.H.F. *et al.*, 2003 *Nucl. Fusion* **43** 1047.
22. Severo J.H.F. *et al.*, 2021 *Plasma Phys. Control. Fusion* **63** 075001.
23. Salewski M. *et al.*, 2018 *Nucl. Fusion* **58** 036017.
24. Johann H., 1931 *Z. Phys.* **69** 185.
25. Bitter M. *et al.*, 1979 *Phys. Rev. Lett.* **42** 304.
26. Bitter M. *et al.*, 1985 *Phys. Rev. A* **32** 3011.
27. Bitter M. *et al.*, 1987 *Plasma Phys. Control. Fusion* **29** 1235.
28. Lieber A.J. *et al.*, 1988 *Phys. Fluids* **31** 687.
29. Bartiromo R. *et al.*, 1989 *Rev. Sci. Instrum.* **60** 237.
30. Platz P. *et al.*, 1993 *20th EPS Conf. on Controlled Fusion and Plasma Physics, Lisbon, 26–30 July 1993* **17C** 1083.
31. von Hamos L., 1933 *Ann. Physik* **17** 716.
32. Källne E. *et al.*, 1985 *Phys. Scr.* **31** 551.
33. Rice J.E. *et al.*, 1995 *Rev. Sci. Instrum.* **66** 752.

J. Rice, *Driven Rotation, Self-Generated Flow, and Momentum Transport in Tokamak Plasmas*, Springer Series on Atomic, Optical, and Plasma Physics 119,
https://doi.org/10.1007/978-3-030-92266-5

34. Rice J.E. *et al.*, 1997 *Nucl. Fusion* **37** 421.
35. Lee W.D. *et al.*, 2003 *Phys. Rev. Lett.* **91** 205003.
36. Rice J.E. *et al.*, 2004 *Nucl. Fusion* **44** 379.
37. Bitter M. *et al.*, 1995 *Rev. Sci. Instrum.* **66** 530.
38. Bitter M. *et al.*, 1999 *Rev. Sci. Instrum.* **70** 292.
39. Ince-Cushman A. *et al.*, 2008 *Rev. Sci. Instrum.* **79** 10E302.
40. Reinke M.L. *et al.*, 2012 *Rev. Sci. Instrum.* **83** 113504.
41. Lee S.G. *et al.*, 2010 *Rev. Sci. Instrum.* **81** 10E506.
42. Pablant N.A. *et al.*, 2012 *Rev. Sci. Instrum.* **83** 083506.
43. Lyu B. *et al.*, 2014 *Rev. Sci. Instrum.* **85** 11E406.
44. Langenberg A. *et al.*, 2018 *Rev. Sci. Instrum.* **89** 10G101.
45. Howell R. *et al.*, 1988 *Rev. Sci. Instrum.* **59** 1521.
46. von Hellermann M. *et al.*, 1995 *Plasma Phys. Control. Fusion* **37** 71.
47. Isler R.C., 1994 *Plasma Phys. Control. Fusion* **36** 171.
48. Zinov́ev V. *et al.*, 1980 *JETP Lett.* **32** 539.
49. Isler R.C. *et al.*, 1981 *Phys. Rev. A* **24** 2701.
50. Fonck R.J. *et al.*, 1983 *Appl. Phys. Lett.* **42** 239.
51. Isler R.C. *et al.*, 1983 *Appl. Phys. Lett.* **42** 355.
52. Groebner R.j. *et al.*, 1983 *Appl. Phys. Lett.* **43** 920.
53. Seraydarian R.P. *et al.*, 1986 *Rev. Sci. Instrum.* **57** 155.
54. Burrell K. *et al.*, 1988 *Nucl. Fusion* **28** 3.
55. Fonck R.J. *et al.*, 1984 *Phys. Rev. A* **29** 3288.
56. Hawkes N.C. and Peacock N.J., 1985 *Nucl. Fusion* **25** 971.
57. Hallock G.A. *et al.*, 1986 *Phys. Rev. Lett.* **56** 1248.
58. Isler R.C. *et al.*, 1986 *Nucl. Fusion* **26** 391.
59. Carolan P.G. *et al.*, 1987 *Phys. Rev. A* **35** 3454.
60. Weisen H. *et al.*, 1989 *Nucl. Fusion* **29** 2187.
61. Ida K. *et al.*, 1991 *Nucl. Fusion* **31** 943.
62. Kim J. *et al.*, 1994 *Phys. Rev. Lett.* **72** 2199.
63. Pütterich T. *et al.*, 2009 *Phys. Rev. Lett.* **102** 025001.
64. Grierson B.A. *et al.*, 2012 *Rev. Sci. Instrum.* **83** 10D529.
65. Haskey S.R. *et al.*, 2018 *Rev. Sci. Instrum.* **89** 10D110.
66. Grierson B.A. *et al.*, 2016 *Rev. Sci. Instrum.* **87** 11E545.
67. Eubank H. *et al.*, 1979 *Proc. 7th Int. Conf. on Plasma Physics and Controlled Nuclear Fusion Research* **1** 167.
68. Snipes J.A. *et al.*, 1990 *Nucl. Fusion* **30** 205.
69. Hutchinson I.H. *et al.*, 2000 *Phys. Rev. Lett.* **84** 3330.
70. Rice J.E. *et al.*, 2000 *Phys. Plasmas* **7** 1825.
71. Rice J.E. *et al.*, 2004 *Phys. Plasmas* **11** 2427.
72. Migliuolo S., 1993 *Nucl. Fusion* **33** 1721.
73. Hutchinson I.H., 1988 *Phys. Rev. A* **37** 4358.
74. LaBombard B. *et al.*, 2004 *Nucl. Fusion* **44** 1047.
75. LaBombard B. *et al.*, 2005 *Phys. Plasmas* **12** 056111.
76. Müller S.H. *et al.*, 2011 *Phys. Rev. Lett.* **106** 115001.
77. Müller S.H. *et al.*, 2011 *Phys. Plasmas* **18** 072504.
78. Hirsch M. *et al.*, 2001 *Plasma Phys. Control. Fusion* **43** 1641.
79. Conway G.D. *et al.*, 2004 *Plasma Phys. Control. Fusion* **46** 951.
80. Conway G.D. *et al.*, 2006 *Nucl. Fusion* **46** S799.
81. Behn R. *et al.*, 1989 *Phys. Rev. Lett.* **62** 2833.
82. Suvorov E.V. *et al.*, 1995 *Plasma Phys. Control. Fusion* **37** 1207.
83. Stejner M. *et al.*, 2013 *Plasma Phys. Control. Fusion* **55** 085002.
84. Stejner M. *et al.*, 2015 *Plasma Phys. Control. Fusion* **57** 062001.
85. Stejner M. *et al.*, 2017 *Plasma Phys. Control. Fusion* **59** 075009.
86. Scott S.D. *et al.*, 1990 *Phys. Rev. Lett.* **64** 531.

87. Kallenbach A. *et al.*, 1991 *Plasma Phys. Control. Fusion* **33** 595.
88. Asakura N. *et al.*, 1993 *Nucl. Fusion* **33** 1165.
89. Nagashima K. *et al.*, 1994 *Nucl. Fusion* **34** 449.
90. Ida K. *et al.*, 1995 *Phys. Rev. Lett.* **74** 1990.
91. Ida K. *et al.*, 1998 *J. Phys. Soc. Japan* **67** 4089.
92. Zastrow K.-D. *et al.*, 1998 *Nucl. Fusion* **38** 257.
93. deGrassie J.S. *et al.*, 2003 *Nucl. Fusion* **43** 142.
94. Nishijima D. *et al.*, 2005 *Plasma Phys. Control. Fusion* **47** 89.
95. deVries P.C. *et al.*, 2006 *Plasma Phys. Control. Fusion* **48** 1693.
96. deGrassie J.S. *et al.*, 2006 *Phys. Plasmas* **13** 112507.
97. Yoshida M. *et al.*, 2007 *Nucl. Fusion* **47** 856.
98. Tala T. *et al.*, 2007 *Nucl. Fusion* **47** 1012.
99. Tala T. *et al.*, 2007 *Plasma Phys. Control. Fusion* **49** B291.
100. Solomon W.M. *et al.*, 2007 *Plasma Phys. Control. Fusion* **49** B313.
101. Yoshida M. *et al.*, 2008 *Phys. Rev. Lett.* **100** 105002.
102. Yoshida M. *et al.*, 2008 *Plasma Fusion Res.* **3** S1007.
103. deVries P.C. *et al.*, 2008 *Nucl. Fusion* **48** 065006.
104. Solomon W.M. *et al.*, 2008 *Phys. Rev. Lett.* **101** 065004.
105. Politzer P.A. *et al.*, 2008 *Nucl. Fusion* **48** 075001.
106. Tala T. *et al.*, 2009 *Phys. Rev. Lett.* **102** 075001.
107. Burrell K. *et al.*, 2009 *Phys. Rev. Lett.* **102** 155003.
108. Yoshida M. *et al.*, 2009 *Fusion Eng. Des.* **84** 2206.
109. Kaye S.M. *et al.*, 2009 *Nucl. Fusion* **49** 045010.
110. Solomon W.M. *et al.*, 2009 *Nucl. Fusion* **49** 085005.
111. Tardini G. *et al.*, 2009 *Nucl. Fusion* **49** 085010.
112. Yoshida M. *et al.*, 2009 *Nucl. Fusion* **49** 115028.
113. Solomon W.M. *et al.*, 2010 *Phys. Plasmas* **17** 056108.
114. Mantica P. *et al.*, 2010 *Phys. Plasmas* **17** 092505.
115. McDermott R.M. *et al.*, 2011 *Plasma Phys. Control. Fusion* **53** 124013.
116. Tala T. *et al.*, 2011 *Nucl. Fusion* **51** 123002.
117. Weisen H. *et al.*, 2012 *Nucl. Fusion* **52** 042001.
118. Weisen H. *et al.*, 2012 *Nucl. Fusion* **52** 114024.
119. Yoshida M. *et al.*, 2012 *Nucl. Fusion* **52** 023024.
120. Yoshida M. *et al.*, 2012 *Nucl. Fusion* **52** 123005.
121. Xu Y. *et al.*, 2013 *Nucl. Fusion* **53** 072001.
122. Ko Won-Ha *et al.*, 2015 *Nucl. Fusion* **55** 083013.
123. Hinton F.L. and Rosenbluth M.N., 1999 *Phys. Lett. A* **259** 267.
124. Goldston R.J. *et al.*, 1981 *J. Comput. Phys.* **43** 61.
125. Pankin A. *et al.*, 2004 *Comput. Phys. Comm.* **159** 157.
126. Challis C.D. *et al.*, 1989 *Nucl. Fusion* **29** 563.
127. Heikkinen J.A. *et al.*, 2001 *J. Comput. Phys.* **173** 527.
128. Salmi A. *et al.*, 2008 *Contrib. Plasma Phys.* **48** 77.
129. Shinohara K. *et al.*, 2006 *J. Plasma Fusion Res.* **1** 007.
130. Wilson J.R. *et al.*, 1998 *Phys. Plasmas* **5** 1721.
131. LeBlanc B.P. *et al.*, 1999 *Phys. Rev. Lett.* **82** 331.
132. Berry L.A. *et al.*, 1999 *Phys. Rev. Lett.* **82** 1871.
133. Jaeger E.F. *et al.*, 2000 *Phys. Plasmas* **7** 3319.
134. Myra J.R. and D'Ippolito D.A. 2002 *Phys. Plasmas* **9** 3867.
135. Wukitch S. *et al.*, 1996 *Phys. Rev. Lett.* **77** 294.
136. Lin Y. *et al.*, 2008 *Phys. Rev. Lett.* **101** 235002.
137. Lin Y. *et al.*, 2009 *Phys. Plasmas* **16** 056102.
138. Lin Y. *et al.*, 2011 *Nucl. Fusion* **51** 063002.
139. Jaeger E.F. *et al.*, 2003 *Phys. Rev. Lett.* **90** 195001.
140. Rice J.E. *et al.*, 2011 *Phys. Rev. Lett.* **106** 215001.

141. Lin Y. *et al.*, 2012 *Plasma Phys. Control. Fusion* **54** 074001.
142. Eriksson L.-G. *et al.*, 1997 *Plasma Phys. Control. Fusion* **39** 27.
143. Eriksson L.-G. *et al.*, 2009 *Plasma Phys. Control. Fusion* **51** 044008.
144. Hedin J. *et al.*, 2002 *Nucl. Fusion* **42** 527.
145. Chang C.S. *et al.*, 1999 *Phys. Plasmas* **6** 1969.
146. Chang C.S. *et al.*, 2000 *Phys. Plasmas* **7** 1089.
147. Perkins F.W. *et al.*, 2001 *Phys. Plasmas* **8** 2181.
148. Chan V.S. *et al.*, 2002 *Phys. Plasmas* **9** 501.
149. Perkins F.W. *et al.*, 2002 *Phys. Plasmas* **9** 511.
150. Eriksson L.-G. and Porcelli F., 2002 *Nucl. Fusion* **42** 959.
151. Murakami S. *et al.*, 2016 *Phys. Plasmas* **23** 012501.
152. Xu Y. *et al.*, 2016 *Nucl. Fusion* **56** 046013.
153. Bilato R. 2017 *Nucl. Fusion* **57** 076017.
154. Eriksson L.-G. *et al.*, 2004 *Phys. Rev. Lett.* **92** 235001.
155. Eriksson L.-G. *et al.*, 1992 *Plasma Phys. Control. Fusion* **34** 863.
156. Rice J.E. *et al.*, 1998 *Nucl. Fusion* **38** 75.
157. Rice J.E. *et al.*, 1999 *Nucl. Fusion* **39** 1175.
158. Hoang G.T. *et al.*, 2000 *Nucl. Fusion* **40** 913.
159. Eriksson L.-G. *et al.*, 2001 *Nucl. Fusion* **41** 91.
160. Rice J.E. *et al.*, 2001 *Nucl. Fusion* **41** 277.
161. Assas S. *et al.*, 2003 'Toroidal plasma rotation in ICRF heated Tore Supra discharges', *30th European Physical Society Conference on Plasma Physics and Controlled Fusion, St. Petersburg, Russia, 7–11 July 2003, ECA Vol.* **27A** P-1.138
162. Noterdaeme J.-M. *et al.*, 2003 *Nucl. Fusion* **43** 274.
163. Rice J.E. *et al.*, 2002 *Nucl. Fusion* **42** 510.
164. Lee J.P. *et al.*, 2011 *Proceedings of the 19th Topical Conf. on Radio Frequency Power in Plasmas* **1406** 459.
165. Peysson Y. and Decker J., 2014 *Fusion Sci. Technol.* **65** 22.
166. Chouli B. *et al.*, 2015 *Plasma Phys. Control. Fusion* **57** 125007.
167. Ince-Cushman A. *et al.*, 2009 *Phys. Rev. Lett.* **102** 035002.
168. Rice J.E. *et al.*, 2009 *Nucl. Fusion* **49** 025004.
169. Rice J.E. *et al.*, 2013 *Nucl. Fusion* **53** 093015.
170. Rice J.E. *et al.*, 2013 *Phys. Rev. Lett.* **111** 125003.
171. Chouli B. *et al.*, 2014 *Plasma Phys. Control. Fusion* **56** 095018.
172. Rice J.E. *et al.*, 2016 *Nucl. Fusion* **56** 036015.
173. Nave M.F.F. *et al.*, 2017 *Nucl. Fusion* **57** 0340002.
174. Koide Y. *et al.*, 1993 *Proc. 14th Int. Conf. on Plasma Physics and Controlled Nuclear Fusion Research* **1** 777.
175. Platz P. *et al.*, 1995 *22nd European Physical Society Conf. on Plasma Physics and Controlled Fusion* **19C** III-337.
176. Nave M.F.F. *et al.*, 2010 *Phys. Rev. Lett.* **105** 105005.
177. Shi Y. *et al.*, 2011 *Phys. Rev. Lett.* **106** 235001.
178. Gao Zhe *et al.*, 2011 *Phys. Plasmas* **18** 082507.
179. Lee Jungpyo *et al.*, 2012 *Plasma Phys. Control. Fusion* **54** 125005.
180. Guan Xiaoyin *et al.*, 2013 *Phys. Plasmas* **20** 022502.
181. Lee J.P. *et al.*, 2014 *Phys. Plasmas* **21** 056106.
182. Lu Z.X. *et al.*, 2015 *Phys. Plasmas* **22** 055705.
183. Lu Z.X. *et al.*, 2015 *Nucl. Fusion* **55** 093012.
184. deGrassie J.S. *et al.*, 2007 *Phys. Plasmas* **14** 056115.
185. Raman R., 2008 *Fusion Eng. Des.* **83** 1368.
186. Onchi T. *et al.*, 2013 *Plasma Phys. Control. Fusion* **55** 035003.
187. Rohollahi A. *et al.*, 2017 *Nucl. Fusion* **57** 056023.
188. Scott S.D. *et al.*, 1985 *Nucl. Fusion* **25** 359.
189. Yoshida M. *et al.*, 2006 *Plasma Phys. Control. Fusion* **48** 1673.

190. deVries P.C. *et al.*, 2008 *Nucl. Fusion* **48** 035007.
191. deVries P.C. *et al.*, 2010 *Plasma Phys. Control. Fusion* **52** 065004.
192. Nave M.F.F. *et al.*, 2012 *Plasma Phys. Control. Fusion* **54** 074006.
193. Romannikov A. *et al.*, 2000 *Nucl. Fusion* **40** 319.
194. Fenzi C. *et al.*, 2011 *Nucl. Fusion* **51** 103038.
195. Crombé K. *et al.*, 2009 *Plasma Phys. Control. Fusion* **51** 055005.
196. deGrassie J.S. *et al.*, 2009 *Nucl. Fusion* **49** 085020.
197. Solomon W.M. *et al.*, 2011 *Nucl. Fusion* **51** 073010.
198. Boedo J.A. *et al.*, 2016 *Phys. Plasmas* **23** 092506.
199. Stacey W.M. *et al.*, 2012 *Phys. Plasmas* **19** 112503.
200. Stacey W.M., 2013 *Nucl. Fusion* **53** 063011.
201. Stacey W.M., 2013 *Phys. Plasmas* **20** 092508.
202. Ou Jing, 2014 *Phys. Plasmas* **21** 072505.
203. Stacey Weston M. and Grierson Brian A., 2014 *Nucl. Fusion* **54** 073021.
204. Seo Janghoon *et al.*, 2014 *Phys. Plasmas* **21** 092501.
205. Pan Chengkang *et al.*, 2014 *Nucl. Fusion* **54** 103003.
206. deGrassie J.S. *et al.*, 2015 *Phys. Plasmas* **22** 080701.
207. Stacey W.M. and Wilks T.M., 2016 *Phys. Plasmas* **23** 012508.
208. Chang C.S. and Ku S., 2008 *Phys. Plasmas* **15** 062510.
209. Stoltzfus-Dueck T., 2012 *Phys. Rev. Lett.* **108** 065002.
210. Parra F.I. *et al.*, 2012 *Phys. Rev. Lett.* **108** 095001.
211. Shaing K.C., 1992 *Phys. Fluids B* **4** 171.
212. Valanju P.M. *et al.*, 1992 *Phys. Fluids B* **4** 2675.
213. Monier-Garbet P. *et al.*, 1997 *Nucl. Fusion* **37** 403.
214. Peeters A.G. *et al.*, 1998 *Plasma Phys. Control. Fusion* **40** 703.
215. Versloot T.W. *et al.*, 2011 *Plasma Phys. Control. Fusion* **53** 065017.
216. Sangwan Deepak *et al.*, 2013 *Phys. Plasmas* **20** 062503.
217. Sangwan Deepak *et al.*, 2015 *Phys. Plasmas* **22** 112501.
218. Omotani J. *et al.*, 2016 *Nucl. Fusion* **56** 124002.
219. Omotani J.T. *et al.*, 2017 *Nucl. Fusion* **57** 066048.
220. Shaing K.C. *et al.*, 1986 *Phys. Fluids* **29** 521.
221. Snipes J.A. *et al.*, 1988 *Nucl. Fusion* **28** 1085.
222. Hender T.C. *et al.*, 1992 *Nucl. Fusion* **32** 2091.
223. La Haye R.J. *et al.*, 1994 *Phys. Plasmas* **1** 373.
224. Smolyakov A.I. *et al.*, 1995 *Phys. Plasmas* **2** 1581.
225. Lazzaro E. *et al.*, 2002 *Phys. Plasmas* **9** 3906.
226. Shaing K.C., 2003 *Phys. Plasmas* **10** 1443.
227. Wolfe S.M. *et al.*, 2005 *Phys. Plasmas* **12** 056110.
228. Zhu W. *et al.*, 2006 *Phys. Rev. Lett.* **96** 225002.
229. Garofalo A.M. *et al.*, 2007 *Nucl. Fusion* **47** 1121.
230. Strait E.J. *et al.*, 2007 *Phys. Plasmas* **14** 056101.
231. Cole A.J. *et al.*, 2007 *Phys. Rev. Lett.* **99** 065001.
232. De Bock M.F.M. *et al.*, 2008 *Nucl. Fusion* **48** 015007.
233. Cole A.J. *et al.*, 2008 *Phys. Plasmas* **15** 056102.
234. Garofalo A.M. *et al.*, 2008 *Phys. Rev. Lett.* **101** 195005.
235. Park J.K. *et al.*, 2009 *Phys. Plasmas* **16** 056115.
236. Garofalo A.M. *et al.*, 2009 *Phys. Plasmas* **16** 056119.
237. Hua M.D. *et al.*, 2010 *Plasma Phys. Control. Fusion* **52** 035009.
238. Sun Y. *et al.*, 2010 *Plasma Phys. Control. Fusion* **52** 105007.
239. Coenen J.W. *et al.*, 2011 *Nucl. Fusion* **51** 063030.
240. Cole A.J. *et al.*, 2011 *Phys. Rev. Lett.* **106** 225002.
241. Callen J.D., 2011 *Nucl. Fusion* **51** 094026.
242. Schaffer M.J. *et al.*, 2011 *Nucl. Fusion* **51** 103028.
243. Sun, Y. *et al.*, 2012 *Nucl. Fusion* **52** 083007.

244. Park J.-K. *et al.*, 2013 *Phys. Rev. Lett.* **111** 095002.
245. Sun, Y. *et al.*, 2013 *Nucl. Fusion* **53** 093010.
246. Piron L. *et al.*, 2013 *Nucl. Fusion* **53** 113022.
247. Rao B. *et al.*, 2013 *Plasma Phys. Control. Fusion* **55** 122001.
248. Honda M. *et al.*, 2014 *Nucl. Fusion* **54** 114005.
249. Shiraki D. *et al.*, 2015 *Plasma Phys. Control. Fusion* **57** 025016.
250. Zhao K.J. *et al.*, 2015 *Nucl. Fusion* **55** 073022.
251. Paz-Soldan C. *et al.*, 2015 *Nucl. Fusion* **55** 083012.
252. Lazzaro E. *et al.*, 2015 *Nucl. Fusion* **55** 093031.
253. Shaing K.C. *et al.*, 2015 *Nucl. Fusion* **55** 125001.
254. Logan N.C. *et al.*, 2016 *Nucl. Fusion* **56** 036008.
255. Goumiri I.R. *et al.*, 2016 *Nucl. Fusion* **56** 036023.
256. Elgriw S. *et al.*, 2016 *Plasma Phys. Control. Fusion* **58** 045002.
257. Liu Yueqiang *et al.*, 2016 *Nucl. Fusion* **56** 056015.
258. Tobias B. *et al.*, 2016 *Phys. Plasmas* **23** 056107.
259. Zhao K.J. *et al.*, 2016 *Nucl. Fusion* **56** 076005.
260. Lee H.H. *et al.*, 2016 *Phys. Plasmas* **23** 082510.
261. Wingen A. *et al.*, 2017 *Nucl. Fusion* **57** 016013.
262. Okabayashi M. *et al.*, 2017 *Nucl. Fusion* **57** 016035.
263. Kim Kimin *et al.*, 2017 *Nucl. Fusion* **57** 036014.
264. Paul E.J. *et al.*, 2017 *Nucl. Fusion* **57** 116044.
265. Honda M. *et al.*, 2017 *Nucl. Fusion* **57** 116050.
266. Yan W. *et al.*, 2018 *Plasma Phys. Control. Fusion* **60** 035007.
267. Jiang M. *et al.*, 2019 *Nucl. Fusion* **59** 046003.
268. Li X.Y. *et al.*, 2019 *Phys. Plasmas* **27** 052512.
269. Yoshida M. *et al.*, 2006 *Plasma Phys. Control. Fusion* **48** A209.
270. Versloot T.W. *et al.*, 2010 *Plasma Phys. Control. Fusion* **52** 045014.
271. Hinton F.L. and Hazeltine R.D., 1976 *Rev. Mod. Phys* **48** 239.
272. Hirshman S.P. and Sigmar D.J., 1981 *Nucl. Fusion* **21** 1079.
273. Kim Y.B. *et al.*, 1991 *Phys. Fluids B* **3** 2050.
274. Newton S. and Helander P., 2006 *Phys. Plasmas* **13** 102505.
275. Wong S.K. *et al.*, 2008 *Phys. Plasmas* **15** 082503.
276. Houlberg W.A. *et al.*, 1997 *Phys. Plasmas* **4** 3230.
277. Pinches S.D. *et al.*, 1998 *Comput. Phys. Commun.* **111** 133.
278. Bergmann A. *et al.*, 2001 *Phys. Plasmas* **8** 5192.
279. Peeters A.G. *et al.*, 2000 *Phys. Plasmas* **7** 268.
280. Belli E.A. *et al.*, 2008 *Plasma Phys. Control. Fusion* **50** 095010.
281. Bell R.E. *et al.*, 1998 *Phys. Rev. Lett.* **81** 1429.
282. Crombé K. *et al.*, 2005 *Phys. Rev. Lett.* **95** 155003.
283. Solomon W.M. *et al.*, 2006 *Phys. Plasmas* **13** 056116.
284. Trier E. *et al.*, 2008 *Nucl. Fusion* **48** 092001.
285. Field A.R. *et al.*, 2009 *Plasma Phys. Control. Fusion* **51** 105002.
286. Bell R.E. *et al.*, 2010 *Phys. Plasmas* **17** 082507.
287. Grierson B.A. *et al.*, 2012 *Phys. Plasmas* **19** 056107.
288. Pütterich T. *et al.*, 2012 *Nucl. Fusion* **52** 083013.
289. Bortolon A. *et al.*, 2013 *Nucl. Fusion* **53** 023002.
290. Grierson B.A. *et al.*, 2013 *Nucl. Fusion* **53** 063010.
291. Viezzer E. *et al.*, 2013 *Plasma Phys. Control. Fusion* **55** 124037.
292. Viezzer E. *et al.*, 2014 *Nucl. Fusion* **54** 012003.
293. Chrystal C. *et al.*, 2014 *Phys. Plasmas* **21** 072504.
294. Burrell K.H. *et al.*, 2014 *Nucl. Fusion* **54** 083020.
295. Lebschy A. *et al.*, 2018 *Nucl. Fusion* **58** 026013.
296. Sarazin Y. *et al.*, 2011 *Nucl. Fusion* **51** 103023.
297. Wang Lu *et al.*, 2016 *Nucl. Fusion* **56** 106017.

298. Stacey W.M. *et al.*, 2006 *Phys. Plasmas* **13** 062508.
299. Bae C. *et al.*, 2013 *Nucl. Fusion* **53** 043011.
300. Haskey S.R. *et al.*, 2018 *Plasma Phys. Control. Fusion* **60** 105001.
301. Claassen H.A. *et al.*, 2000 *Phys. Plasmas* **7** 3699.
302. Rogister A.L. *et al.*, 2002 *Nucl. Fusion* **42** 1144.
303. deGrassie J.S. *et al.*, 2004 *Phys. Plasmas* **11** 4323.
304. Churchill R.M. *et al.*, 2013 *Rev. Sci. Instrum.* **84** 093505.
305. Theiler C. *et al.*, 2014 *Nucl. Fusion* **54** 083017.
306. Viezzer E. *et al.*, 2015 *Nucl. Fusion* **55** 123002.
307. Bae C. *et al.*, 2014 *Phys. Plasmas* **21** 012504.
308. McDermott R.M. *et al.*, 2009 *Phys. Plasmas* **16** 056103.
309. Whyte D.G. *et al.*, 2010 *Nucl. Fusion* **50** 105005.
310. Hubbard A.E. *et al.*, 2011 *Phys. Plasmas* **18** 056115.
311. White A.E. *et al.*, 2011 *Nucl. Fusion* **51** 113005.
312. Hubbard A.E. *et al.*, 2012 *Nucl. Fusion* **52** 114009.
313. Hughes J.W. *et al.*, 2013 *Nucl. Fusion* **53** 043016.
314. Cziegler I. *et al.*, 2013 *Phys. Plasmas* **20** 055904.
315. Walk J.R. *et al.*, 2014 *Phys. Plasmas* **21** 056103.
316. Rice J.E. *et al.*, 2015 *Nucl. Fusion* **55** 033014.
317. Rice J.E. *et al.*, 2017 *Nucl. Fusion* **57** 116004.
318. Coffey I.H., Barnsley R., Keenan F.P. *et al.*, 1996 in *Proceedings of the 11th Colloquium on UV and X-ray Spectroscopy of Astrophysical and Laboratory Plasmas, Nagoya, Japan, 1995, p.431, Frontiers Science Series* No.15 (Editors: K. Yamashita and T. Watanabe), Universal Academy Press, Tokyo, Japan
319. Rice J.E. *et al.*, 2007 *Fusion Sci. Technol.* **51** 288.
320. Rice J.E. *et al.*, 2008 *Plasma Phys. Control. Fusion* **50** 124042.
321. Park Jong-Kyu *et al.*, 2013 *Nucl. Fusion* **53** 063012.
322. Rice J.E. *et al.*, 2007 *Nucl. Fusion* **47** 1618.
323. Lyu B. *et al.*, 2020 *Phys. Plasmas* **27** 022511.
324. Sakamoto Y. *et al.*, 2006 *Plasma Phys. Control. Fusion* **48** A63.
325. Kosuga Y. *et al.*, 2010 *Phys. Plasmas* **17** 102313.
326. Gürcan Ö.D. *et al.*, 2007 *Phys. Plasmas* **14** 042306.
327. Diamond P.H. *et al.*, 2008 *Phys. Plasmas* **15** 012303.
328. Gürcan Ö.D. *et al.*, 2010 *Phys. Plasmas* **17** 112309.
329. Polevoi A.R. *et al.*, 2017 *Nucl. Fusion* **57** 022014.
330. Sorbom B.N. *et al.*, 2014 *Fusion Eng. Des.* **100** 378.
331. Porte L. *et al.*, 2007 *Nucl. Fusion* **47** 952.
332. Yoshida M. *et al.*, 2009 *Phys. Rev. Lett.* **103** 065003.
333. deGrassie J. *et al.*, 'ECH and core intrinsic rotation reversal in DIII-D' *Proceedings of the 21st Topical Conference on Radio Frequency Power in Plasmas, 2015 California USA* Vol. 1689, AIP Publishing 090001.
334. deGrassie J. *et al.*, 2016 *Phys. Plasmas* **23** 082501.
335. McDermott R.M. *et al.*, 2011 *Plasma Phys. Control. Fusion* **53** 035007.
336. Seol J. *et al.*, 2012 *Phys. Rev. Lett.* **109** 195003.
337. Yoshida M. *et al.*, 2013 *Nucl. Fusion* **53** 083022.
338. Shi Y.J. *et al.*, 2013 *Nucl. Fusion* **53** 113031.
339. Yoshida M. *et al.*, 2015 *Nucl. Fusion* **55** 073014.
340. Shi Y.J. *et al.*, 2016 *Nucl. Fusion* **56** 016014.
341. Grierson B.A. *et al.*, 2017 *Phys. Rev. Lett.* **118** 015002.
342. Wang W.X. *et al.*, 2006 *Phys. Plasmas* **13** 092505.
343. Wang W.X. *et al.*, 2017 *Phys. Plasmas* **24** 092501.
344. Hughes J.W. *et al.*, 2010 *Nucl. Fusion* **50** 064001.
345. Chrystal C. *et al.*, 2017 *Phys. Plasmas* **24** 042501.
346. Chrystal C. *et al.*, 2017 *Phys. Plasmas* **24** 056113.

347. Chrystal C. *et al.*, 2020 *Nucl. Fusion* **60** 036003.
348. Ernst D.R. *et al.*, 1998 *Phys. Plasmas* **5** 665.
349. Sakamoto Y. *et al.*, 2001 *Nucl. Fusion* **41** 865.
350. Rice J.E. *et al.*, 2003 *Nucl. Fusion* **43** 781.
351. Fiore C.L. *et al.*, 2010 *Nucl. Fusion* **50** 064008.
352. Sips A.C.C. *et al.*, 1998 *Plasma Phys. Control. Fusion* **40** 1171.
353. Burrell K.H. *et al.*, 1998 *Plasma Phys. Control. Fusion* **40** 1585.
354. Doyle E.J. *et al.*, 2000 *Plasma Phys. Control. Fusion* **42** A237.
355. Tala T.J.J. *et al.*, 2001 *Plasma Phys. Control. Fusion* **43** 507.
356. Ide S. *et al.*, 2004 *Nucl. Fusion* **44** 87.
357. Shaing K.C., 2001 *Phys. Rev. Lett.* **86** 640.
358. Coppi B., 2002 *Nucl. Fusion* **42** 1.
359. Wang W.X. *et al.*, 2009 *Phys. Rev. Lett.* **102** 035005.
360. Wang W.X. *et al.*, 2010 *Phys. Plasmas* **17** 072511.
361. Wang W.X. *et al.*, 2011 *Phys. Rev. Lett.* **106** 085001.
362. Wang W.X. *et al.*, 2011 *Phys. Plasmas* **18** 042502.
363. Kwon J.M. *et al.*, 2012 *Nucl. Fusion* **52** 013004.
364. McDevitt C.J. *et al.*, 2009 *Phys. Rev. Lett.* **103** 205003.
365. Spineanu Florin and Vlad Madalina, 2012 *Nucl. Fusion* **52** 114019.
366. Wang Lu and Diamond P.H. *et al.*, 2013 *Phys. Rev. Lett.* **110** 265006.
367. Peng Shuitao *et al.*, 2017 *Nucl. Fusion* **57** 036003.
368. T.Tala *et al.*, 2016 *Proceedings 43rd EPS Conference on Plasma Physics* O2.104.
369. Rice J.E. *et al.*, 2021 *Nucl. Fusion* **61** 026013.
370. Rodriguez-Fernandez P. *et al.*, 2020 *J. Plasma Phys.* **86** 865860503.
371. LaBombard B. *et al.*, 2008 *Phys. Plasmas* **15** 056106.
372. Rice J.E. *et al.*, 2005 *Nucl. Fusion* **45** 251.
373. Duval B.P. *et al.*, 2007 *Plasma Phys. Control. Fusion* **49** B195.
374. Duval B.P. *et al.*, 2008 *Phys. Plasmas* **15** 056113.
375. Camenen Y. *et al.*, 2010 *Plasma Phys. Control. Fusion* **52** 124037.
376. Camenen Y. *et al.*, 2010 *Phys. Rev. Lett.* **105** 135003.
377. Camenen Y. *et al.*, 2011 *Nucl. Fusion* **51** 073039.
378. Ball Justin *et al.*, 2018 *Nucl. Fusion* **58** 026003.
379. Stoltzfus-Dueck T. *et al.*, 2015 *Phys. Rev. Lett.* **114** 245001.
380. Stoltzfus-Dueck T. *et al.*, 2015 *Phys. Plasmas* **22** 056118.
381. Stoltzfus-Dueck T. *et al.*, 2012 *Phys. Plasmas* **22** 055908.
382. Bortolon A. *et al.*, 2006 *Phys. Rev. Lett.* **97** 235003.
383. Rice J.E. *et al.*, 2011 *Nucl. Fusion* **51** 083005.
384. Rice J.E. *et al.*, 2011 *Phys. Rev. Lett.* **107** 265001.
385. Rice J.E. *et al.*, 2012 *Phys. Plasmas* **19** 056106.
386. Rice J.E. *et al.*, 2013 *Nucl. Fusion* **53** 033004.
387. Reinke M.L. *et al.*, 2013 *Nucl. Fusion* **53** 043006.
388. McDermott R.M. *et al.*, 2014 *Nucl. Fusion* **54** 043009.
389. Gao C. *et al.*, 2014 *Nucl. Fusion* **54** 083025.
390. Hillesheim J.C. *et al.*, 2015 *Nucl. Fusion* **55** 032003.
391. Na D.H. *et al.*, 2017 *Nucl. Fusion* **57** 126008.
392. Grierson B.A. *et al.*, 2019 *Phys. Plasmas* **26** 042304.
393. Shukla G. *et al.*, 2019 *Nucl. Fusion* **59** 106049.
394. Bernardo J. *et al.*, 2015 *Plasma Phys. Control. Fusion* **57** 035002.
395. Na D.H. *et al.*, 2016 *Nucl. Fusion* **56** 036011.
396. Trier E. *et al.*, 2017 *Nucl. Fusion* **57** 046021.
397. Ejima S. *et al.*, 1982 *Nucl. Fusion* **22** 1627.
398. Y.Shimomura, N.Suzuki, M.Sugihara, T.Tsuda, K.Odajima and T.Tsunematsu, 'Empirical Scaling of Energy Confinement Time of L-Mode and Optimized Mode and Some Consideration of Reactor Core Plasma in Tokamak' JAERI Report 87-080 (1987)

399. Arnichand H. *et al.*, 2014 *Nucl. Fusion* **54** 123017.
400. Ennever P. *et al.*, 2015 *Phys. Plasmas* **22** 072507.
401. Angioni C. *et al.*, 2011 *Phys. Rev. Lett.* **107** 215003.
402. Lee J.A. *et al.*, 2018 *Phys. Plasmas* **25** 022513.
403. Shi Y.J. *et al.*, 2017 *Nucl. Fusion* **57** 066040.
404. Barnes M. *et al.*, 2013 *Phys. Rev. Lett.* **111** 055005.
405. Scarabosio A. *et al.*, 2006 *Plasma Phys. Control. Fusion* **48** 663.
406. Rice J.E. *et al.*, 2020 *Nucl. Fusion* **60** 105001.
407. http://www.jt60sa.org/b/n1/introduction.htm
408. Albanese R. *et al.*, 2017 *Nucl. Fusion* **57** 016010.
409. Watterson R.L. *et al.*, 1985 *Phys. Fluids* **28** 2857.
410. Brower D.L. *et al.*, 1987 *Phys. Rev. Lett.* **59** 48.
411. Rettig C.L. *et al.*, 2001 *Phys. Plasmas* **8** 2232.
412. Sung C. *et al.*, 2013 *Nucl. Fusion* **53** 083010.
413. Krämer-Flecken A. *et al.*, 2015 *New J. Phys.* **17** 073007.
414. Arnichand H. *et al.*, 2015 *Nucl. Fusion* **55** 093021.
415. Angioni C. *et al.*, 2005 *Phys. Plasmas* **12** 040701.
416. Lin L. *et al.*, 2009 *Plasma Phys. Control. Fusion* **51** 065006.
417. Cao N. *et al.*, 2019 *Nucl. Fusion* **59** 104001.
418. Camenen Y. *et al.*, 2017 *Plasma Phys. Control. Fusion* **59** 034001.
419. Chilenski M.A. *et al.*, 2017 *Nucl. Fusion* **57** 126013.
420. Reinke, M.L. *et al.*, 2013 *Plasma Phys. Control. Fusion* **55** 012001.
421. Howard N.T. *et al.*, 2014 *Plasma Phys. Control. Fusion* **56** 124004.
422. Idomura Yasuhiro, 2017 *Phys. Plasmas* **24** 080701.
423. Wang Lu *et al.*, 2016 *Phys. Plasmas* **23** 042309.
424. Hornsby W.A. *et al.*, 2018 *Nucl. Fusion* **58** 056008.
425. Fedorczak N. *et al.*, 2012 *Nucl. Fusion* **52** 103013.
426. Hariri F. *et al.*, 2016 *Phys. Plasmas* **23** 052512.
427. Barnes M. and Parra F.I., 2019 *Plasma Phys. Control. Fusion* **61** 025003.
428. Lee Jungpyo *et al.*, 2015 *Plasma Phys. Control. Fusion* **57** 125006.
429. Hornsby W.A. *et al.*, 2017 *Nucl. Fusion* **57** 046008.
430. Stoltzfus-Dueck T., 2019 *Plasma Phys. Control. Fusion* **61** 124003.
431. Connor J.W. *et al.*, 1987 *Plasma Phys. Control. Fusion* **29** 919.
432. Guttenfelder W. *et al.*, 2017 *Nucl. Fusion* **57** 056022.
433. Peeters A.G. *et al.*, 2007 *Phys. Rev. Lett.* **98** 265003.
434. Mattor N. and Diamond P.H., 1988 *Phys. Fluids* **31** 1180.
435. Peeters A.G and Angioni C., 2005 *Phys. Plasmas* **12** 072515.
436. Peeters A.G. *et al.*, 2006 *Plasma Phys. Control. Fusion* **48** B413.
437. Shaing K.C., 2001 *Phys. Plasmas* **8** 193.
438. Hahm T.S. *et al.*, 2007 *Phys. Plasmas* **14** 072302.
439. Waltz R. *et al.*, 2007 *Phys. Plasmas* **14** 122507.
440. Gürcan Ö.D. *et al.*, 2008 *Phys. Rev. Lett.* **100** 135001.
441. Hahm T.S. *et al.*, 2008 *Phys. Plasmas* **15** 055902.
442. Peeters A.G. *et al.*, 2009 *Phys. Plasmas* **16** 042310.
443. Yoon E.S. and Hahm T.S., 2010 *Nucl. Fusion* **50** 064006.
444. Zimmermann C.F.B. *et al.*, 2021 'Analysis and modeling of momentum transport based on NBI modulation experiments at ASDEX Upgrade'
445. Callen J.D. *et al.*, 2009 *Phys. Plasmas* **16** 082504.
446. Howard N.T. *et al.*, 2016 *Nucl. Fusion* **56** 014004.
447. Malkov M.A. *et al.*, 2015 *Phys. Plasmas* **22** 032506.

Index

J. Rice, *Driven Rotation, Self-Generated Flow, and Momentum Transport in Tokamak Plasmas*, Springer Series on Atomic, Optical, and Plasma Physics 119,
https://doi.org/10.1007/978-3-030-92266-5